BRITISH RAILWA`

LOCO|

CW00344966

THIRTY-THIRD EDITION – SUMMER AUTUMN 1994

The Complete Guide to all
BR-operated Locomotives

Peter Fox and
Richard Bolsover

PLATFORM
5

ISBN 1 872524 64 8

CONTENTS

FROM THE PUBLISHER

This book, formerly known as 'Motive Power Pocket Book', contains details of all BR-operated locomotives. This includes BR and SNCF locomotives which will work through the Channel Tunnel, but the 'Eurostar' trains owned by the state railway companies of Britain, France and Belgium and the stock owned by the Channel Tunnel Company is included in Railways Pocket Book No. 3 – DMUs and Channel Tunnel Stock. We would like to thank all who have helped by contributing information or photographs.

Platform 5 Publishing are still the only company to provide the enthusiast or transport official with complete information on BR rolling stock. This book is, at the time of publication, the most up-to-date publication of its type on the market. Information is updated to 27th June 1994.

In the centre of this issue will be found details of our book club which features pre-publication discounts on most Platform 5 books (excluding our BR pocket books). It is hoped that enthusiasts will find this facility convenient and cost effective.

We regret that we have had to increase the price mid-year, contrary to our normal policy. This is due to massive rises in paper costs.

On page 92, readers interested in European railways will find details of our new magazine 'Today's Railways' – the first such magazine of its kind.

NOTES

The following notes are applicable to locomotives:

LOCOMOTIVE CLASS DETAILS

Principal details and dimensions are given for each class in metric units. Imperial equivalents are also given for power. Maximum speeds are still quoted in miles per hour since the operating department of BR still uses imperial units. Since the present maximum permissible speed of certain classes of locomotives is different from the design speed, these are now shown separately in class details. In some cases certain low speed limits are arbitrary and may occasionally be raised raised when necessary if a locomotive has to be pressed into passenger service.

Standard abbreviations used are:

ABB	ABB Transportation Ltd.	kN	kilonewtons
BR	British Railways	kV	kilovolts
BREL	British Rail Engineering Ltd. (later BREL Ltd.)	kW	kilowatts
BRML	British Rail Maintenance Ltd.	lbf	pounds force
hp	horse power	mph	miles per hour
km/h	kilometres/hour	RA	Route availability
		t	tonnes

LOCOMOTIVE DETAIL DIFFERENCES

Detail differences which affect the areas and types of train which locos work are shown. Where detail differences occur within a class or part class of locos., these are shown against the individual loco number. Except where shown, diesel locomotives have no train heating equipment. Electric or electro-diesel locomotives are assumed to have train heating unless shown otherwise. Standard abbreviations used are:

a	Train air brakes only
c	Cab to shore radio-telephone fitted.
e	Fitted with electric heating apparatus (ETH).
r	Fitted with radio electronic token block equipment.
s	Slow speed control fitted (and operable).
t	Fitted with automatic vehicle identification transponders.
v	Train vacuum brakes only.
x	Dual train brakes (air & vacuum).
y	ETH equipped but equipment isolated.
+	Extended range locos with Additional fuel tank capacity compared with others in class.

NAMES AND ALLOCATIONS

All official names are shown as they appear on the locomotive i.e. all upper case or upper & lower case lettering.

(S) denotes stored serviceable and (U) stored unserviceable. Last known allocations of stored locomotives are shown, but readers should note that locomotives may not necessarily be stored at their home depots.

After the locomotive number are shown any notes regarding braking, heating etc., the livery code (in bold type), the pool code, the depot code and name if any. Locomotives which have been renumbered in recent years show the last number in parentheses after the current number. Where only a few locomotives in a class are named, these are shown in a separate table at the end of the class or sub-class.

Thus the layout is as follows:

No.	Old No.	Notes	Liv.	Pool		Depot	Name
47777	(47636)	+	**RX**	PXLB		CD	Restored

GENERAL INFORMATION ON BRITISH RAILWAYS' LOCOMOTIVES

CLASSIFICATION & NUMBERING

Initially BR diesel locomotives were allocated numbers in the 1xxxx series, with electrics allotted numbers in the 2xxxx series. Around 1957 diesel locomotives were allocated new four digit numbers with 'D' prefixes. Diesel electric shunters in the 13xxx series had the '1' replaced by a 'D', but diesel mechanical shunters were completely renumbered. Electric locomotives retained their previous numbers but with an 'E' prefix. When all standard gauge steam locomotives had been withdrawn, the prefix letter was removed.

In 1972, the present TOPS numbering system was introduced whereby the loco number consisted of a two-digit class number followed by a serial number. In some cases the last two digits of the former number were generally retained (classes 20, 37, 50), but in other classes this is not the case. In this book former TOPS numbers carried by converted locos. are shown in parentheses. Full renumbering information is to be found in the 'Diesel & Electric loco Register', the third edition of which is now available.

Diesel locomotives are classified as "types" depending on their engine horsepower as follows:

Type	Engine hp.	Old Number Range		Current Classes
1	800 – 1000	D 8000 – D 8999		20
2	1001 – 1499	D 5000 – D 6499/D 7500 – D 7999		26, 31.
3	1500 – 1999	D 6500 – D 7499		33, 37.
4	2000 – 2999	D 1 – D 1999		47, 50.
5	3000 +	D 9000 – D 9499		56, 58, 59, 60.
Shunter	Under 300	D 2000 – D 2999		03.
Shunter	300 – 799	D 3000 – D 4999		08, 09.

Class 14 (650 hp diesel hydraulics) were numbered in the D95xx series.

Electric and electro-diesel locomotives are classified according to their supply system. Locomotives operating on a d.c. system are allocated classes 71 – 80, whilst a.c. or dual voltage locomotives start at Class starting at 81.

Departmental locomotives which remain self propelled or which are likely to move around on a day to day basis are classified 97.

WHEEL ARRANGEMENT

For main line diesel and electric locomotives the system whereby the number of driven axles on a bogie or frame is denoted by a letter (A = 1, B = 2, C = 3 etc.) and the number of undriven axles is noted by a number is used. The letter 'o' after a letter indicates that each axle is individually powered and a + sign indicates that the bogies are intercoupled.

For shunters and steam locomotives the Whyte notation is used. The number of leading wheels are given, followed by the number of driving wheels and then the trailing wheels. Suffix 'T' on a steam locomotive indicates a tank locomotive, and 'PT' a pannier tank loco.

HAULING CAPABILITY OF DIESEL LOCOS

The hauling capability of a diesel locomotive depends basically upon three factors:

1. Its adhesive weight. The greater the weight on its driving wheels, the greater the adhesion and thus more tractive power can be applied before wheel slip occurs.

2. The characteristics of its transmission. In order to start a train the locomotive has to exert a pull at standstill. A direct drive diesel engine cannot do this, hence the need for transmission. This may be mechanical, hydraulic or electric. The current BR standard for locomotives is electric transmission. Here the diesel engine drives a generator or alternator and the current produced is fed to the traction motors. The force produced by each driven wheel depends on the current in its traction motor. In other words the larger the current, the harder it pulls.

As the locomotive speed increases, the current in the traction motors falls hence the *Maximum Tractive Effort* is the maximum force at its wheels that the locomotive can exert at a standstill. The electrical equipment cannot take such high currrents for long without overheating. Hence the *Continuous Tractive Effort* is quoted which represents the current which the equipment can take continuously.

3. The power of its engine. Not all of this power reaches the rail as electrical machines are approximately 90% efficient. As the electrical energy passes through two such machines (the generator/alternator and the traction motors), the *Power At Rail* is about 81% (90% of 90%) of the engine power, less a further amount used for auxiliary equipment such as radiator fans, traction motor cooling fans, air compressors, battery charging, cab heating, ETH, etc. The power of the locomotive is proportional to the tractive effort times the speed. Hence when on full power there is a speed corresponding to the continuous tractive effort.

HAULING CAPABILITY OF ELECTRIC LOCOS

Unlike a diesel locomotive, an electric locomotive does not develop its power on board and its performance is determined only by two factors, namely its weight and the characteristics of its electrical equipment. Whereas a diesel locomotive tends to be a constant power machine, the power of an electric locomotive varies considerably. Up to a certain speed it can produce virtually a constant tractive effort. Hence power rises with speed according to the formula given in section 3 above, until a maximum speed is reached at which tractive effort falls, such that the power also falls. Hence the power at the speed corresponding to the maximum tractive effort is lower than the maximum.

BRAKE FORCE

The brake force is a measure of the braking power of a locomotive. This is shown on the locomotive data panels so that railway staff can ensure that sufficient brake power is available on freight trains.

TRAIN HEATING EQUIPMENT

Electric train heating (ETH) is now the standard system in use on BR for loco-hauled trains. Locomotives which were equipped to provide steam heating have had this equipment removed or rendered inoperable (isolated). Electric heat is provided from the locomotive by means of a separate alternator on the loco., except in the case of and classes 33 and 50 which have a d.c. generator. The *ETH Index* is a measure of the electrical power available for train heating. All electrically heated coaches have an ETH index and the total of these in a train must not exceed the ETH power of a locomotive.

ROUTE AVAILABILITY

This is a measure of a railway vehicle's axle load. The higher the axle load of a vehicle, the higher the RA number on a scale 1 to 10. Each route on BR has an RA number and in theory no vehicle with a higher RA number may travel on that route without special clearance. Exceptions are made, however.

MULTIPLE AND PUSH-PULL WORKING

Multiple working between diesel locomotives on BR has usually been provided by means of an electro-pneumatic system, with special jumper cables connecting the locos. A coloured symbol is painted on the end of the locomotive to denote which system is in use. The original Class 47/7s used a time-division multiplex (t.d.m.) system which utilised the existing RCH (an abbreviation for the former railway clearing house, a pre-nationalisation standards organisation) jumper cables for push-pull working. These had in the past only been used for train lighting control, and more recently for public address (pa) and driver – guard communication. A new standard t.d.m. system is now fitted to all a.c. electric locomotives and other vehicles, enabling them to work in both push-pull and multiple working modes.

BR DIESEL LOCOMOTIVES

CLASS 03   BR SHUNTER   0-6-0

Built: 1960 at BR Doncaster Works.
Engine: Gardner 8L3 of 152 kW (204 hp) at 1200 rpm.
Transmission: Mechanical. Fluidrive type 23 hydraulic coupling to Wilson-Drewry CA5R7 gearbox with SCG type RF11 final drive.
Max. Tractive Effort: 68 kN (15300 lbf).
Brake Force: 13 t.
Weight: 31 t.
Max. Speed: 28 mph.
Length over Buffers: 7.92 m.
Wheel Diameter: 1092 mm.
RA: 1.

Formerly numbered 2079.

03079 v   HZSH   RY

CLASS 08   BR SHUNTER   0-6-0

Built: 1953 – 62 by BR at Crewe, Darlington, Derby, Doncaster or Horwich Works.
Engine: English Electric 6KT of 298 kW (400 hp) at 680 rpm.
Main Generator: English Electric 801.
Traction Motors: Two English Electric 506.
Max. Tractive Effort: 156 kN (35000 lbf).
Cont. Tractive Effort: 49 kN (11100 lbf) at 8.8 mph.
Power At Rail: 194 kW (260 hp).
Brake Force: 19 t.
Design Speed: 20 mph.
Max. Speed: 15 or 20* mph.
Length over Buffers: 8.92 m.
Wheel Diameter: 1372 mm.
Weight: 50 t.
RA: 5.

Non standard liveries:

08414 is 'D' with RfD brandings and also carries its former number D 3529.
08500 is red lined out in black & white.
08519/730/867 are BR black.
08593 is Great Eastern blue lined out in red and also carries its former number D 3760.
08601 is London Midland & Scottish Railway black.
08629 is Royal purple.
08642 is London & South Western Railway black and also carries its former number D 3809.
08689 is 'D' with Railfreight general brandings.
08715 is in experimental dayglo orange livery.
08721 is blue with a red & yellow stripe ("Red Star" livery).
08883 is Caledonian blue.
08907 is London & North Western Railway black.
08933 is 'D' but with two orange cabside stripes.
08938 is grey and red.

n – Waterproofed for working at Oxley Carriage Depot.

z – Fitted with buckeye adaptor at nose end for HST depot shunting.
§ – Fitted with yellow flashing light and siren for working between Ipswich Yard and Cliff Quay.

Formerly numbered in series 3000 – 4192. 08600 was numbered 97800 whilst in departmental use between 1979 and 1989.

CLASS 08/0. Standard Design.

08388	a		FP	FDSI	IM		08516	a	D	FDSK	KY
08389	a			DATI	TI		08517	a		EWSX	SF (S)
08397	a		F	LWSP	SP		08519	a	0	LBBY	BY
08401	a		D	FDSI	IM		08523	x		EWOC	OC
08402	a		D	PXLS	CD		08525	x	F	HISL	NL
08405	a		D	FDSI	IM		08526	x		EWSF	SF
08410	a		D	HJSA	BR		08527	x	D	KCSI	IL
08411	a			LGML	ML		08528	x	D	ENSN	TO
08413	a		D	DASY	TI		08529	x		ENSN	TO
08414	a*§	0		EWSF	SF		08530	x	D	DAWE	TI
08415	x			LWSP	SP		08531	x	D	DAWE	TI
08417	a		D	CDJD	DY		08534	x	D	LWCL	CL
08418	a		F	FDSD	DR		08535	x	D	DASY	TI
08428	a			LBBS	BS		08536	x		HISE	DY
08441	a			ENSN	TO		08538	x	D	ENSN	TO
08442	a		F	FDSK	KY		08540	x	D	ENSN	TO
08445	a			FDSX	IM		08541	x	D	EWSF	SF
08447	a			LWCX	CL (U)		08542	x	F	EWSF	SF
08448	a			LBBX	BS (S)		08543	x	D	LBBS	BS
08449	a			ENSN	TO		08561	x		LGMX	ML
08451	x			HFSN	WN		08567	x		LBBY	BY
08454	x			HFSN	WN		08568	x		KGSS	ZH
08460	a		F	EWOC	OC		08569	x		DAAN	AN
08466	a		FO	FDSI	IM		08571	xz		HBSH	EC
08472	a			HBSH	BN		08573	x		KCSI	IL
08480	az		G	EWOC	OC		08575	x	BS	DATI	TI
08481	x			LNCF	CF		08576	x		LNCF	CF
08482	a		D	DAAN	AN		08577	x		FMSY	TE
08483	a		D	HJSA	BR		08578	x	R	PXLS	HT
08484	a		D	KWSW	ZN		08580	x		ENSN	TO
08485	a			LWSP	SP		08581	x	BS	FDSD	DR
08489	a		F	LWSP	SP		08582	a	D	FMSY	TE
08492	a			ENSN	TO		08585	x		DAAN	AN
08493	a			LNCF	CF		08586	a	F	LGAY	AY
08495	x			ENSN	TO		08587	x		FMSY	TE
08499	a		F	FDSK	KY		08588	xz	BS	HISL	NL
08500	x		0	FDSD	DR		08593	x	0	EWSF	SF
08506	a			LGML	ML		08594	x		PXLS	CA
08509	a		F	FDSD	DR		08597	x		ENSN	TO
08510	a			FDSD	DR		08599	x		PXLS	CD
08511	a			ENSN	TO		08600	a	D	EWEH	EH
08512	a		F	FDSD	DR		08601	x	0	LBBS	BS
08514	a			FDSD	DR		08605	x		FDSK	KY

08607	x		ENSN	TO	08701	x	R	PXLS	HT
08610	x		LBBX	BS	08702	x		PXLS	WN
08611	x		HFSL	LO (S)	08703	a		DAAN	AN
08616	x		HGSS	TS	08706	x		FDSK	KY
08617	x		HFSN	WN	08709	x		EWSF	SF
08619	x		LWSX	SP (S)	08711	x		PXLS	CA
08622	x		LGML	ML	08713	a		FDSD	DR
08623	x		LBBS	BS	08714	x	RX	PXLS	CA
08624	x		DAAN	AN	08715	v	0	EWSF	SF
08625	x		LBBY	BY	08718	x		LGML	ML
08628	x		LBBY	BY	08720	x	D	LGML	ML
08629	x	0	KWSW	ZN	08721	x	0	HFSL	LO
08630	x		LGML	ML	08723	x		ENSN	TO
08632	x		FDSI	IM	08724	x		HBSN	BN
08633	x	RX	PXLS	CD	08730	x	0	KGSS	ZH
08635	x		PXLS	CD	08731	x		LGML	ML
08641	xz	D	HJSL	LA	08734	x		LBBS	BS
08642	x*	0	DAWE	TI	08735	x		LGML	ML
08643	xz	D	HJSA	BR	08737	x	F	DAAN	AN
08644	xz	I	HJSL	LA	08738	x	D	LGML	ML
08645	xz	D	HJSL	LA	08739	x		DAAN	AN
08646	x	F	EWOC	OC	08740	x	F	EWSF	SF
08648	x*	D	HJSL	LA	08742	x		PXLS	CD
08649	x	G	KESE	ZG	08745	xz	BS	DATI	TI
08651	x	F	EWOC	OC	08746	x	D	LBBS	BS
08653	x*		DADR	TI	08750	x		EWSF	SF
08655	x*	F	DAWE	TI	08751	x	FE	DASY	TI
08661	a		DAYX	TI (U)	08752	x	C	EWSF	SF
08662	x		FDSK	KY	08754	x		HASS	IS
08663	a	D	HJSL	LA	08756	x	D	LNCF	CF
08664	x		EWOC	OC	08757	x	RX	PXLS	CA
08665	x		FDSI	IM	08758	x		EWSF	SF
08666	x		LWSX	SP (S)	08762	x		HASS	IS
08668	x		PXLS	CD	08765	xn	D	LBBS	BS
08670	a		EWSF	SF	08768	x		LWCL	CL
08673	x	10	DAYX	TI (U)	08770	a	D	LNCF	CF
08675	x	F	LGAY	AY	08773	x		ENSX	TO (U)
08676	x		LWSP	SP	08775	x		EWSF	SF
08682	x		KDSD	ZF	08776	a	D	FDSK	KY
08683	x		LBBY	BY	08780	x		HJSE	LE
08685	x		PXLS	CA	08782	a		FDSK	KY
08689	a	0	EWSF	SF	08783	x		FDSK	KY
08690	x		HISE	DY	08784	x		DAAN	AN
08691	x	G	DAYX	TI (U)	08786	a	D	LNCF	CF
08693	x		LGML	ML	08790	x		HFSL	LO
08694	x		DAAN	AN	08792	x		LNBZ	BZ
08695	x		PXLS	CD	08795	x	D	HJSE	LE
08696	a	D	HFSN	WN	08798	x		LNCF	CF
08697	x		HISE	DY	08799	x		DAAN	AN
08698	a		EWSU	SU	08801	x		LNCF	CF
08700	a		EWSX	SF (S)	08804	x		PXLS	CD (U)

08805 x	FO	HGSS	TS
08806 a	F	FMSY	TE
08807 x	BS	LBBY	BY
08810 a		HSSN	NC
08811 a*		EWSX	SF (S)
08813 x	D	FDSD	DR
08815 x		LWSP	SP
08817 x	BS	LWSP	SP
08818 x		PXLS	CD
08819 x	D	LNBZ	BZ
08822 x		HJSE	LE (U)
08823 a		KDSD	ZF
08824 a	F	FDSD	DR
08825 a		DADR	TI
08826 a		LWCL	CL
08827 a		LWCL	CL
08828 a		EWSF	SF
08830 x*		HLSV	CF
08834 x	FD	HBSH	BN
08836 x		HJSA	BR (U)
08837 x*	D	DADR	TI
08842 x		DAWE	TI
08844 x		LWCX	CL (U)
08847 x*		KESE	ZG
08853 xr		HBSH	EC
08854 x*		EWEH	EH
08856 x		DAAN	AN
08865 x		PXLS	CA
08866 x		EWSF	SF
08867 x	0	LWSP	SP
08869 x	G	HSSN	NC
08872 x	D	DAAN	AN
08873 x	M	PXLS	WN
08877 x	D	FDSD	DR
08878 x		EWSX	SF (U)
08879 x		DATI	TI
08881 x	D	LGAY	AY
08882 x		LGML	ML
08883 x	0	LGML	ML
08884 x		LWSP	SP
08886 x		PXLS	HT
08887 x		HFSN	WN
08888 xz	R	PXLS	HT
08890 x	D	PXLS	WN
08891 x		DAAN	AN
08892 x*	D	DAWE	TI
08893 x	D	LBBX	BS (U)
08894 x		LWSP	SP
08896 x		PXLS	CD
08897 x	D	PXLS	CD
08899 x		HISE	DY
08900 x	D	LWSP	SP
08901 xn		LBBX	BS (U)
08902 x		DAAN	AN
08903 x		FDSD	DR
08904 x		EWOC	OC
08905 x		DASY	TI
08906 x		LGAY	AY
08907 x	0	DAAN	AN
08908 xz		HISL	NL
08909 x		EWSF	SF
08910 x		LWCL	CL
08911 x	D	LWCL	CL
08912 x		LWCL	CL
08913 x	D	DADR	TI
08914 x		LBBS	BS
08915 x	F	LWSP	SP
08918 x	D	LWSP	SP
08919 x		PXLS	CD
08920 x	F	LBBS	BS
08921 x		PXLS	CD
08922 x	D	LGML	ML
08924 x	D	EWOC	OC
08925 x		LWSP	SP
08926 x		DAWE	TI
08927 x		LBBY	BY
08928 x	FR	HSSN	NC
08931 x		FMSY	TE
08932 x		LNCF	CF
08933 x*	0	EWEH	EH
08934 x		HFSN	WN
08938 xr	0	LGMX	ML
08939 x		DAAN	AN
08940 x		EWEH	EH
08941 x		LNCF	CF
08942 x		LNCF	CF
08944 x	D	EWOC	OC
08946 x	D	DASY	TI
08947 x		EWOC	OC
08948 x		GPSS	OC
08950 x	I	HISL	NL
08951 x	D	DAAN	AN
08952 x		LGML	ML
08953 x	D	LNCF	CF
08954 x	F	LNBZ	BZ
08955 x		LNBZ	BZ
08956 x		CDJD	DY
08957 x		EWSF	SF
08958 x		EWSX	SF (U)

Names:

08578 Libert Dickinson	08757 EAGLE C.U.R.C.
08633 The Sorter	08869 The Canary
08649 G.H. Stratton	08888 Postman's Pride
08701 GATESHEAD TMD 1852 – 1991	08950 Neville Hill 1st

Class 08/9. Fitted with cut-down cab and headlight for Cwmmawr branch.

08993	(08592)	x	LNCF	CF	ASHBURNHAM
08994	(08462)	a	**FR** LNCX	CF	
08995	(08687)	a	**FC** LNCF	CF	KIDWELLY

CLASS 09   BR SHUNTER   0-6-0

Built: 1959 – 62 by BR at Darlington or Horwich Works.
Engine: English Electric 6KT of 298 kW (400 hp) at 680 rpm.
Main Generator: English Electric 801.
Traction Motors: English Electric 506.
Max. Tractive Effort: 111 kN (25000 lbf).
Cont. Tractive Effort: 39 kN (8800 lbf) at 11.6 mph.
Power At Rail: 201 kW (269 hp).
Brake Force: 19 t.
Weight: 50 t.
Max. Speed: 27 mph.
Length over Buffers: 8.92 m.
Wheel Diameter: 1372 mm.
RA: 5.

Class 09/0 were formerly numbered 3665 – 71, 3719 – 21, 4099 – 4114.

CLASS 09/0. Built as Class 09.

09001		LNCX	CF	09014	**D**	FDSK	KY
09003		EWSU	SU	09015	**D**	LNCF	CF
09004		HWSU	SU	09016	**D**	EWSU	SU
09005	**D**	FDSK	KY	09018		EWSU	SU
09006		EWSU	SU	09019	**D**	EWSU	SU
09007		EWSU	SU	09020		EWSU	SU
09008	**D**	LNCF	CF	09021		DAWE	TI
09009	**D**	EWSU	SU	09022		DAWE	TI
09010	**D**	EWSU	SU	09023		EWSU	SU
09011	**D**	DAWE	TI	09024	**D**	EWSU	SU
09012	**D**	EWSU	SU	09025		HWSU	SU
09013	**D**	LNCF	CF	09026	**D**	HWSU	SU

Names:

09009 Three Bridges C.E.D.	09026 William Pearson
09012 Dick Hardy	

CLASS 09/1. Converted from Class 08. 110 V electrical equipment.

09101	(08833)	**D**	EWOC	OC
09102	(08832)	**D**	EWOC	OC
09103	(08766)	**D**	LGML	ML
09104	(08749)	**D**	LBBS	BS
09105	(08835)	**D**	LNCF	CF

| 09106 | (08759) | **D** | FMSY | TE |
| 09107 | (08845) | **D** | LNCF | CF |

CLASS 09/2. Converted from Class 08. 90 V electrical equipment.

09201	(08421)	**D**	ENSN	TO
09202	(08732)	**D**	LGML	ML
09203	(08781)	**D**	LNCF	CF
09204	(08717)	**D**	FMSY	TE
09205	(08620)	**D**	LGML	ML

CLASS 20   ENGLISH ELECTRIC TYPE 1   Bo–Bo

Built: 1957 – 68 by English Electric Company at Vulcan Foundry, Newton le Willows or Robert Stephenson & Hawthorn, Darlington. 20001 – 128 were originally built with disc indicators whilst 20129 – 228 were built with four character headcode panels.
Engine: English Electric 8SVT Mk. II of 746 kW (1000 hp) at 850 rpm.
Main Generator: English Electric 819/3C.
Traction Motors: English Electric 526/5D (20001 – 48) or 526/8D (others).
Max. Tractive Effort: 187 kN (42000 lbf).
Cont. Tractive Effort: 111 kN (25000 lbf) at 11 mph.
Power At Rail: 574 kW (770 hp).   **Length over Buffers:** 14.25 m.
Brake Force: 35 t.   **Wheel Diameter:** 1092 mm.
Design Speed: 75 mph.   **Weight:** 73.5 t.
Max. Speed: 60 mph.   **RA:** 5.
Train Brakes: Air & vacuum.
Multiple Working: Blue Star Coupling Code.

Formerly numbered in series 8007 – 8190, 8315 – 8325.

CLASS 20/0. BR-owned Locomotives.

20007	st		TAKX	KI (S)	
20016	st		LBHB	BS (U)	
20032	s		TAKX	KI (S)	
20057	st		LBHB	BS (U)	
20059	st	**FR**	LBHB	BS (U)	
20066			LBHB	BS (S)	
20072	st		TAKX	KI (U)	
20075	st	**T**	TAKB	BS	Sir William Cooke
20081	st		LBHB	BS (U)	
20087	st	**BS**	LBHB	BS (S)	
20092		**CS**	LBHB	BS	
20104	st	**FR**	TAKX	KI (S)	
20117	st		TAKX	KI (U)	
20118		**FR**	LBHB	BS	
20121	st		TAKX	KI (U)	
20128	st		TAKB	BS	
20131	st	**T**	TAKB	BS	Almon B. Strowger
20132	st	**FR**	LBHB	BS	
20138		**FR**	LBHB	BS (S)	
20165		**FR**	LBHB	BS	

20168	st		LBHB	BS (U)	
20169	st	CS	LBHB	BS	
20187	st	T	TAKB	BS	Sir Charles Wheatstone
20190	st		TAKX	KI (U)	
20215	st	FR	TAKX	KI (U)	

CLASS 20/9. Privately-owned by Hunslet – Barclay Ltd.

Used on weedkilling trains.
Non-standard Livery: Hunslet – Barclay two-tone grey with red lettering.

20901	(20041)	t	0	XYPD	ZK	NANCY
20902	(20060)		0	XYPD	ZK	LORNA
20903	(20083)		0	XYPD	ZK	ALISON
20904	(20101)		0	XYPD	ZK	JANIS
20905	(20225)	t	0	XYPD	ZK	IONA
20906	(20219)		0	XYPD	ZK	Kilmarnock 400

CLASS 31   BRUSH TYPE 2   A1A – A1A

Built: 1957 – 62 by Brush Traction at Loughborough.
31102/5 – 7/10/25/34/44/418/44/50/61 retain two headcode lights. Others
have roof-mounted headcode boxes.
Engine: English Electric 12SVT of 1100 kW (1470 hp) at 850 rpm.
Main Generator: Brush TG160-48.
Traction Motors: Brush TM73-68.
Max. Tractive Effort: 160 kN (35900 lbf) (190 kN (42800 lbf)*).
Cont. Tractive Effort: 83 kN (18700 lbf) at 23.5 mph. (99 kN (22250 lbf) at
19.7 mph *.)

Power At Rail: 872 kW (1170 hp).	**Length over Buffers:** 17.30 m.
Brake Force: 49 t.	**Driving Wheel Diameter:** 1092 mm.
Design Speed: 90 (80*) mph.	**Centre Wheel Diameter:** 1003 mm.
Max. Speed: 60 mph (90 mph 31/4)	**Weight:** 107 – 111 t.
RA: 5 or 6.	**ETH Index (Class 31/4):** 66.

Train Brakes: Air & vacuum.
Multiple Working: Blue Star Coupling Code.
Communication Equipment: This class is in the process of being fitted with cab
to shore radio-telephone.

Non standard liveries:

31116 is red, yellow, red and grey with 'Infrastructure' branding.
31413 is BR blue with yellow cabsides, a light blue stripe along the bottom of
the body and a red band around the bottom of the cabs.

Formerly numbered 5520 – 5699, 5800 – 5862 (not in order).

CLASS 31/1. Standard Design. RA5.

31102		C	LBDB	BS	31113	C	LBDB	BS
31105	*	C	LBDB	BS	31116	0	ENPN	TO
31106	*	C	LBDB	BS	31119	C	LWDC	SP
31107		C	LBDB	BS	31125	C	LBDB	BS
31110		C	LBDB	BS	31126	C	LWDC	SP
31112	*	C	LBDB	BS	31128	FO	LBRB	BS

31130	FC	LWNC	SP		31207	C	LWNC	SP
31132	FO	LBRB	BS		31209	FA	ENXX	TO (S)
31134	C	LWNC	SP		31219	C	ENRN	TO
31135	C	ENPN	TO		31224	C	LWNC	SP
31142	C	LWDC	SP		31229	C	LWNC	SP
31144	C	LWDC	SP		31230 *	FO	ENRN	TO
31145	C	LWNC	SP		31232	C	LBRB	BS
31146 r	C	LBDB	BS		31233	C	LWDC	SP
31147 r	C	LBDB	BS		31235	C	LWDC	SP
31149	FR	ENRN	TO		31237	C	LBDB	BS
31154	C	LWDC	SP		31238	C	LWDC	SP
31155	FA	LBDB	BS		31242	C	LWRC	SP
31158	C	LNXX	BS		31247	FR	ENRN	TO
31159	C	LWDC	SP		31248	FO	LNXX	BS (U)
31160	F	LWRC	SP		31250	C	ENPN	TO
31163	C	LWNC	SP		31252	FO	ENXX	TO (S)
31164	FO	LBRB	BS		31255	C	LWNC	SP
31165	G	ENRN	TO		31263	C	LWDC	SP
31166	C	LBDB	BS		31268	C	ENRN	TO
31171	FO	LNXX	BS		31270	FC	LWRC	SP
31174	C	LNXX	BS		31271	FA	ENPN	TO
31178	C	LBDB	BS		31272	C	LWRC	SP
31180	FR	ENRN	TO		31273	C	LBDB	BS
31181	C	ENRN	TO		31275	FC	LWNC	SP
31184	FO	ENXX	TO (U)		31276	FC	ENPN	TO
31185	C	LBDB	BS		31285	C	LWDC	SP
31186	C	ENRN	TO		31290	C	ENRN	TO
31187	C	ENRN	TO		31294	FA	ENRN	TO
31188	C	LWNC	SP		31301	FR	LWRC	SP
31190	C	LWDC	SP		31302	FP	LWNC	SP
31191	C	ENRN	TO		31304	FC	LWNC	SP
31199	FC	LWNC	SP		31306	C	LWDC	SP
31200	FC	LWNC	SP		31308	C	ENPN	TO
31201	FC	LWNC	SP		31312	FC	LWNC	SP
31203	C	LWDC	SP		31317	FO	LBRB	BS
31205	FR	ENRN	TO		31319	FC	LWNC	SP
31206	C	LBRB	BS		31327	FR	LWDC	SP

Names:

31102 Cricklewood	31130 Calder Hall Power Station
31105 Bescot TMD	31146 Brush Veteran
31106 The Blackcountryman	31147 Floreat Salopia
31107 John H Carless VC	31165 Stratford Major Depot
31116 RAIL Celebrity	31233 Severn Valley Railway

CLASS 31/4. Equipped with Train Heating. RA6.
CLASS 31/5. Dedicated for Civil Engineer's Department Use. Train Heating Equipment isolated. RA6.

31403		ENDN	TO	
31405	M	LBDB	BS	Mappa Mundi
31407 (31507)	M	ENDN	TO	

31408		LWDC	SP
31410	**RR** LWKC	SP	Granada Telethon
31411 (31511)	**D** LBRB	BS	
31512 (31412)	**C** LWDC	SP	
31413	**O** LNXX	BS (U)	
31514 (31414)	**C** LBRB	BS	
31415		LBRB	BS
31516 (31416)	**C** LNXX	BS	
31417	**D** LBDB	BS	
31418		LWRC	SP
31519 (31419)	**C** LWDC	SP	
31420 (31172)	**M** LBDB	BS	
31421 (31140)	**RR** LWKC	SP	Wigan Pier
31422 (31522)	**M** LBDB	BS	
31423 (31197)	**M** LBDB	BS	Jerome K. Jerome
31524 (31424)	**C** LBDB	BS	
31526 (31426)	**C** LBRB	BS	
31427 (31194)		LWDC	SP
31530 (31430)	**C** LBDB	BS	Sister Dora
31531 (31431)	**C** ENPN	TO	
31432 (31153)		LWKC	SP
31533 (31433)	**C** LBRB	BS	
31434 (31258)		LBDB	BS
31435 (31179)	**C** LBDB	BS	
31537 (31437)	**C** LBDB	BS	
31538 (31438)		LWRC	SP
31439 (31239)	**RR** LWKC	SP	North Yorkshire Moors Railway
31541 (31441)	**C** ENPN	TO	
31444 (31544)	**C** LWKC	SP	Keighley and Worth Valley Railway
31545 (31445)		LBDB	BS
31546 (31446)	**C** LBDB	BS	
31547 (31447)	**C** ENRN	TO	
31548 (31448)	**C** LBRB	BS	
31549 (31449)	**C** ENPN	TO	
31450 (31133)		LBDB	BS
31551 (31451)	**C** ENPN	TO	
31552 (31452)	**C** ENPN	TO	
31553 (31453)	**C** ENRN	TO	
31554 (31454)	**C** LBDB	BS	
31455 (31555)	**RR** LWKC	SP	'Our Eli'
31556 (31456)	**C** LWDC	SP	
31558 (31458)	**C** ENPN	TO	Nene Valley Railway
31459 (31256)		ENDN	TO
31461 (31129)	**D** ENDN	TO	
31462 (31315)	**D** LBDB	BS	
31563 (31463)	**C** ENRN	TO	
31465 (31565)	**RR** LWKC	SP	
31466 (31115)	**C** ENRN	TO	
31467 (31216)		LBDB	BS
31468 (31568)	**C** LBDB	BS	The Enginemen's Fund

31569 (31469)   C   ENRN   TO

CLASS 33   BRCW TYPE 3   Bo – Bo

Built: 1960 – 62 by the Birmingham Railway Carriage & Wagon Company, Smethwick.
Engine: Sulzer 8LDA28 of 1160 kW (1550 hp) at 750 rpm.
Main Generator: Crompton Parkinson CG391B1.
Traction Motors: Crompton Parkinson C171C2.
Max. Tractive Effort: 200 kN (45000 lbf).
Cont. Tractive Effort: 116 kN (26000 lbf) at 17.5 mph.
Power At Rail: 906 kW (1215 hp).   **Length over Buffers:** 15.47 m.
Brake Force: 35 t.   **Wheel Diameter:** 1092 mm.
Design Speed: 85 mph.   **Weight:** 77.5 t (78.5 t Class 33/1).
Max. Speed: 60 mph.   **RA:** 6.
Train Heating: Electric (y isolated). **ETH Index:** 48.
Train Brakes: Air & vacuum.
Multiple Working: Blue Star Coupling Code.
Communication Equipment: This class is in the process of being fitted with cab to shore radio-telephone.

Formerly numbered in series 6500 – 97 but not in order.
Note: 33116 carries its original number D 6535.

Class 33/0. Standard Design.

33002	y	C	EWDB	SL	Sea King
33008	y	G	EWDB	SL	Eastleigh
33012	e		EWDB	SL	
33019	e	C	EWDB	SL	Griffon
33021	e	FA	EWRB	SL	
33023	e		EWDB	SL	
33025	e	C	EWDB	SL	Sultan
33026	e	C	EWDB	SL	Seafire
33029	e		EWDB	SL	
33030	e	C	EWDB	SL	
33035	y	N	EWDB	SL	Spitfire
33042	e	FA	EWRB	SL	
33046	y	C	EWDB	SL	Merlin
33048	es		EWRB	SL	
33051	e	C	EWDB	SL	Shakespeare Cliff
33052	e		EWRB	SL	Ashford
33053	e	FA	EWRB	SL	
33057	ys	C	EWDB	SL	Seagull
33063	ys	FA	EWRB	SL	
33064	e	FA	ENZX	SL	
33065	e	C	EWDB	SL	Sealion

Class 33/1. Fitted with Buckeye Couplings & SR Multiple Working Equipment for use with SR EMUs, TC stock & class 73.

Also fitted with flashing light adaptor for use on Weymouth Quay line.

33103	e	C	EWRB	SL	

▲ Faded BR blue liveried Class 03 No. 03079 at Sandown, IoW on 17th July 1993.   *M Hilbert*

▼ Class 08 No. 08723 is pictured shunting HEA wagons at Sandiacre, near Toton on 24th March 1993.   *Brian Denton*

▲ Departmental grey liveried Class 09 No. 09201 at Wetmore, near Burton-on-Trent on 7th August 1993. *Hugh Ballantyne*

▼ Central Services liveried Class 20 No. 20169 is seen stabled at Stratford on 30th April 1994. *Kevin Conkey*

Class 20s Nos. 20131 'Almon B. Strowger' and 20118, the former in BR Telecoms livery and the latter in Railfreight red-stripe livery, pass Chelvey with a railtour from London Paddington on 2nd May 1994.

Nic Joynson

Class 31s Nos. 31132 and 31125, both in old Railfreight livery, pass Greenholme whilst heading a welded rail train on 1st September 1990.

Hugh Ballantyne

Civil-link liveried Class 33 No. 33202 'The Burma Star' passes through Wandsworth Road, South London with a southbound engineer's train on 13th April 1994.

Kevin Conkey

▲ Class 37s Nos. 37128, in revised blue livery, and 37235, in Railfreight livery with no sub-sector markings, pass through Camden Road station on 17th October 1992 with a Pengam – Felixstowe Freightliner. *Kevin Conkey*

▼ The 16.48 Holyhead – Manchester Victoria passes Winwick Junction on 21st May 1994 headed by Regional Railways liveried Class 37 No. 37418 'East Lancashire Railways'. *Les Nixon*

Class 43 No. 43189 leads the 08.47 Penzance – Paddington at Shaldon Bridge, Teignmouth on 10th March 1994.

Colin J Marsden

Parcels red liveried Class 47 No. 47582 'County of Norfolk' passes 56064 in the down loop at Clay Cross on 28th August 1992 whilst working the 12.44 Newcastle – Plymouth Rail Express Systems service.

Paul D Shannon

33109	e	D	EWDB	SL	Captain Bill Smith RNR
33116	e		EWDB	SL	Hertfordshire Rail Tours
33117	e		EWRB	SL	

Class 33/2. Built to Former Loading Gauge of Hastings Line.

33202	ys	C	EWDB	SL	The Burma Star
33204	es	FD	EWDB	SL	
33206	es	FD	EWDB	SL	
33207	ys	FA	EWDB	SL	Earl Mountbatten of Burma
33208	es	C	EWDB	SL	

CLASS 37   ENGLISH ELECTRIC TYPE 3   Co–Co

Built: 1960 – 5 by English Electric Company at Vulcan Foundry, Newton le Willows or Robert Stephenson & Hawthorn, Darlington. 37003 – 116/350/1/9 with the exception of 37019*/047/053/065*/072*/073/074/075*/100* (* one end only) retain box-type route indicators, the remainder having central headcode panels/marker lamps.
Engine: English Electric 12CSVT of 1300 kW (1750 hp) at 850 rpm.
Main Generator: English Electric 822/10G.
Traction Motors: English Electric 538/A.
Max. Tractive Effort: 245 kN (55500 lbf).
Cont. Tractive Effort: 156 kN (35000 lbf) at 13.6 mph.
Power At Rail: 932 kW (1250 hp).   **Length over Buffers:** 18.75 m.
Brake Force: 50 t.   **Wheel Diameter:** 1092 mm.
Design Speed: 90 mph.   **Weight:** 103 – 108 t.
Max. Speed: 80 mph.   **RA:** 5 or 7.
Train Heating: Electric (Class 37/4 only). **ETH Index:** 38
Train Brakes: Air & vacuum.
Multiple Working: Blue Star Coupling Code.
Communication Equipment: This class is in the process of being fitted with cab to shore radio-telephone.

a Vacuum brake isolated.

Formerly numbered 6600 – 8, 6700 – 6999 (not in order). 37271 – 4 are the second locos to carry these numbers. They were renumbered to avoid confusion with Class 37/3 locos.

Class 37/0. Unrefurbished Locos. Technical details as above. RA5.
Note: 37009 is shown officially as 37340, but had not been renumbered at the time of writing.

37003	+	C	FDDI	IM	
37004		FM	LNXX	ML (U)	
37009	+	FD	FDDI	IM	
37010		C	EWCN	TO	
37012		C	EWCN	TO	
37013	+	F	EWDS	SF	
37015	+	FD	FMRY	TE	
37019	+	FD	FMRY	TE	
37023		C	EWDS	SF	Stratford TMD Quality Approved
37025		BR	LGSV	ML	Inverness TMD Quality Assured

Number		Class	Depot	Region	Name
37026 (37320)	+	FD	LWCC	CD	Shapfell
37035		C	EWCN	TO	
37037 (37321)		FM	LNDK	CF	
37038		C	EWCN	TO	
37040		FM	EWCN	TO	
37042	+	FM	EWCN	TO	
37043 (37354)		C	LGBM	ML	
37045 (37355)	+	F	FMRY	TE	
37046		C	EWCN	TO	
37047	+	FD	EWDS	SF	
37048		FM	EWCN	TO	
37049		C	FDDI	IM	Imperial
37051		FM	LGBM	ML	
37053	+	FD	FMDY	TE	
37054		C	EWDS	SF	
37055	+	FD	EWDS	SF	
37057	+	BR	ENDN	TO	
37058	+	C	FDDI	IM	
37059	+	FD	FMRY	TE	Port of Tilbury
37063	+	FD	FMRY	TE	
37065	+	FD	EWRN	TO	
37066	+	C	LGBM	ML	
37068 (37356)	+	FD	FMDY	TE	Grainflow
37069	+	C	LGBM	ML	
37070		FD	EWRN	TO	
37071	+	C	LGBM	ML	
37072	+	D	EWCN	TO	
37073	+	FD	LGBM	ML	Fort William/An Gearasdan
37074	+	FD	EWRN	TO	
37075	+	F	FMDY	TE	
37077		FM	EWDS	SF	
37078	+	FM	LNXX	ML (U)	
37079 (37357)	+	FD	FDRI	IM	
37080		FP	LGBM	ML	
37083	+	C	FDDI	IM (U)	
37087		C	LGSV	ML	
37088 (37323)		C	LGBM	ML	Clydesdale
37092		C	ENDN	TO	
37095	+	C	FDDI	IM	
37097		C	EWCN	TO	
37098	+	C	EWCN	TO	
37099 (37324)		C	LGSV	ML	Clydebridge
37100	+	FM	LGBM	ML	
37101	+	FD	FDYX	IM (U)	
37104		C	FDYX	IM (U)	
37106	+	C	EWDS	SF	
37107	+	FD	LWCC	CD	
37108 (37325)	+	F	LWCC	CD	
37109		FM	EWDS	SF	
37110	+	FD	DAYX	TI (U)	
37111 (37326)		FM	LGBM	ML	Glengarnock

37113	+	**FD** LGBM	ML	Radio Highland
37114	+	**C** ENDN	TO	City of Worcester
37116	+	**BR** LGBM	ML	
37128	+	**BR** FMRY	TE	
37131	+	**FD** DAMT	TI	
37133		**C** LGBM	ML	
37137 (37312)		**FM** EWCN	TO	Clyde Iron
37138		**FM** EWRN	TO	
37139	+	**FC** FDRI	IM	
37140		**C** EWDS	SF	
37141		**C** LNDK	CF	
37142		**C** LNDK	CF	
37144	r	**FA** FDYX	IM (U)	
37146		**C** LNDK	CF	
37152 (37310)		**I** LGSV	ML	
37153		**C** LGBM	ML	
37154	+	**FD** LGBM	ML (S)	
37156 (37311) r		**C** LGSV	ML	British Steel Hunterston
37158		**C** LNDK	CF	
37162	+	**D** EWCN	TO	
37165 (37374)	+	**C** LGBM	ML	
37167	+	**FC** EWDS	SF	
37170	r	**C** LGSV	ML	
37174		**C** EWCN	TO	
37175		**C** LGBM	ML	
37178	+	**FD** DAMT	TI	
37184		**C** LGBM	ML	
37185	+	**C** ENDN	TO	Lea & Perrins
37188		**C** LGBM	ML	
37191		**C** LNDK	CF	
37194	+	**FD** EWRB	SL	British International Freight Association
37196		**C** LGBM	ML	
37197	+	**C** LNDK	CF	
37198	+	**C** EWDB	SL	
37201		**C** LGBM	ML (U)	Saint Margaret
37202		**FM** FMRY	TE	
37203		**FM** EWCN	TO	
37207		**C** LNDK	CF	
37209		**BR** FDYX	IM (U)	
37211		**C** LGBM	ML	
37212	+	**FC** LGBM	ML	
37213	+	**FC** EWCN	TO	
37214	+	**FA** LGBM	ML	
37216	r +	**G** EWDS	SF	Great Eastern
37217	+	FMRY	TE	
37218	+	**FD** DAYX	TI (S)	
37219	r	EWDS	SF	
37220	+	**FP** EWRB	SL	
37221		**I** LGSV	ML	
37222	+	**FC** EWCN	TO	

37223		+	FC	FDRI	IM	
37225		+	FD	DAMT	TI	
37227		+	FM	EWCN	TO	
37229		+	FC	LNLK	CF	
37230		+	C	LNDK	CF	
37232	r		C	LGBM	ML	The Institution of Railway Signal Engineers
37235		+	F	FDRI	IM	
37238		+	FD	DAYX	TI (U)	
37239		+	FC	FMRY	TE	The Coal Merchants' Association of Scotland
37240		+	C	LGBM	ML	
37241			FM	EWDS	SF	
37242		+	FD	EWDS	SF	
37244		+	FD	EWDS	SF	
37245			C	EWRB	SL	
37248		+	FM	LNXX	CF (U)	
37250		+	FM	LGSV	ML	
37251		+	I	LGSV	ML	The Northern Lights
37252			FD	FDRI	IM	
37254		+	C	LNDK	CF	
37255		+	C	LGSV	ML	
37258		+	C	LNDK	CF	
37261		+	FD	LGBM	ML	Caithness
37262		+	D	LGBM	ML	Dounreay
37263			C	LNDK	CF	
37264			C	EWCN	TO	
37271 (37303)		+	FD	FDRI	IM	
37272 (37304)		+	FD	FMRY	TE	
37274 (37308)		+	C	EWDB	SL	
37275		+		LGBM	ML	Oor Wullie
37278		+	FC	ENXX	TO (U)	
37280		+	FP	EWDS	SF	
37285		+	F	FMRY	TE	
37293		+	FM	EWRB	SL	
37294		+	C	LGBM	ML	
37298		+	FD	DAMT	TI	

Class 37/3. Unrefurbished locos fitted with regeared (CP7) bogies.
Details as Class 37/0 except:
Max. Tractive Effort: 250 kN (56180 lbf).
Cont. Tractive Effort: 184 kN (41250 lbf) at 11.4 mph.

37340						
37350 (37119)	+	FP	FMDY	TE	
37351 (37002)	+	C	LGPM	ML	
37358 (37091)		F	FMDY	TE	
37359 (37118)		FP	FMDY	TE	
37370 (37127)		C	EWDS	SF	
37371 (37147)	+	C	EWDS	SF	
37372 (37159)		C	EWRN	TO	
37375 (37193)	+	C	EWDB	SL	

```
37376 (37199) +  FC  EWDS     SF
37377 (37200) +  C   EWDB     SL
37378 (37204) +  FD  FMDY     TE
37379 (37226)    C   EWDS     SF    Ipswich WRD Quality Assured
37380 (37259)    FC  EWRB     SL
37381 (37284) +  FD  FDYX     IM (U)
37382 (37145)    FP  FDYX     IM (U)
```

Class 37/4. Refurbished locos fitted with train heating. Main generator replaced by alternator. Regeared (CP7) bogies. Details as class 37/0 except:

Main Alternator: Brush BA1005A.
Max. Tractive Effort: 256 kN (57440 lbf).
Cont. Tractive Effort: 184 kN (41250 lbf) at 11.4 mph.
Power At Rail: 935 kW (1254 hp).
All have twin fuel tanks.

```
37401 (37268) r  FD  LGHM     ML    Mary Queen of Scots
37402 (37274) r  M   LWMC     CD    Bont Y Bermo
37403 (37307) r  G   LGHM     ML    Ben Cruachan
37404 (37286) r  F   LGHM     ML
37405 (37282) r  M   LWCC     CD    Strathclyde Region
37406 (37295) r  FD  LGHM     ML    The Saltire Society
37407 (37305) r  M   LWMC     CD    Loch Long
37408 (37289)    BR  LWMC     CD    Loch Rannoch
37409 (37270) r  F   LGHM     ML    Loch Awe
37410 (37273) r  M   LGHM     ML    Aluminium 100
37411 (37290)    FD  LNXX     CF
37412 (37301)    F   LNLK     CF
37413 (37276) r  FD  LNLK     CF    Loch Eil Outward Bound
37414 (37287) r  RR  LWMC     CD    Cathays C&W Works
                                    1846 – 1993
37415 (37277) r  M   FMCY     TE
37416 (37302) r  M   LNLK     CF
37417 (37269) r  F   LWCC     CD    Highland Region
37418 (37271) r  RR  LWMC     CD    East Lancashire Railway
37419 (37291) r  M   FMCY     TE
37420 (37297) r  M   LWCC     CD    The Scottish Hosteller
37421 (37267) r  RR  LWMC     CD    The Kingsman
37422 (37266) r  RR  LWMC     CD    Robert F. Fairlie Locomotive
                                    Engineer 1831 – 1885
37423 (37296) r  FD  LGHM     ML    Sir Murray Morrison
              1873 – 1948 Pioneer of British Aluminium Industry
37424 (37279) r  M   LGHM     ML    Isle of Mull
37425 (37292) r  FA  LWMC     CD    Sir Robert McAlpine/
                                    Concrete Bob (opp. sides)
37426 (37299) r  M   FMCY     TE
37427 (37288) r  RR  LGPV     ML    Highland Enterprise
37428 (37281) r  FP  LGPV     ML    David Lloyd George
37429 (37300) r  RR  LWMC     CD    Eisteddfod Genedlaethol
37430 (37265) r  M   LGHM     ML    Cwmbrân
37431 (37272) r  M   LGPV     ML
```

Class 37/5. Refurbished locos. Main generator replaced by alternator. Regeared (CP7) bogies. Details as class 37/4 except:

Max. Tractive Effort: 248 kN (55590 lbf).
All have twin fuel tanks.

37501	(37005)		**FM** FEPS	IM (S)	
37502	(37082)		**FM** FEPS	IM (U)	
37503	(37017)		**FM** LWCC	CD	
37504	(37039)		**FM** FEPS	IM (S)	
37505	(37028)		**I** LGSV	ML	
37506	(37007)		**FM** FEPS	TE (U)	British Steel Skinningrove
37507	(37036)		**FM** FEPS	IM (U)	
37508	(37090)	s	**FM** FEPS	IM (S)	
37509	(37093)		**F** LWCC	CD	
37510	(37112)		**I** LGSV	ML	
37511	(37103)		**FM** FEPS	IM (U)	Stockton Haulage
37512	(37022)		**FM** FDCI	IM	Thornaby Demon
37513	(37056)		**FM** FEPS	IM (U)	
37514	(37115)	s	**FM** FEPS	TE	
37515	(37064)	s	**FM** FDCI	IM	
37516	(37086)	s	**FM** FMCY	TE	
37517	(37018)	as	**FM** FDCI	IM	
37518	(37076)		**FM** LWCC	CD	
37519	(37027)		**FM** FDCI	IM	
37520	(37041)		**FM** LWCC	CD	
37521	(37117)		**FP** LNLK	CF	
37667	(37151)	as	**FP** ESPS	SF	
37668	(37257)	s	**FP** LNLK	CF	
37669	(37129)		**FD** LNXX	CF	
37670	(37182)		**FD** LNLK	CF	St. Blazey T&RS Depot
37671	(37247)		**FD** LNLK	CF	Tre Pol and Pen
37672	(37189)	s	**FD** LNLK	CF	Freight Transport Association
37673	(37132)		**FD** LNLK	CF	
37674	(37169)		**FD** LNLK	CF	
37675	(37164)	s	**FD** LWCC	CD	
37676	(37126)		**FA** ESPS	SF	
37677	(37121)		**F** FDCI	IM	
37678	(37256)		**FA** ESPS	SF	
37679	(37123)		**FA** ESPS	SF	
37680	(37224)		**FA** FDCI	IM	
37682	(37236)		**FA** FMCY	TE	
37683	(37187)		**I** LGSV	ML	
37684	(37134)		**FA** FDCI	IM	Peak National Park
37685	(37234)		**I** LGSV	ML	
37686	(37172)		**FA** LWCC	CD	
37687	(37181)		**FA** FEPS	IM (U)	
37688	(37205)		**FA** FDCI	IM	Great Rocks
37689	(37195)	s	**F** FDCI	IM	
37690	(37171)		**FO** FEPS	IM (U)	
37691	(37179)	s	**FO** FEPS	IM (S)	
37692	(37122)	s	**FC** LGPM	ML	The Lass O' Ballochmyle

37693	(37210)	s	**FC** LGPM	ML
37694	(37192)	s	**FC** FDCI	IM
37695	(37157)	s	**FC** LNLK	CF
37696	(37228)	s	**FC** LGPM	ML
37697	(37243)	s	**FC** FMCY	TE
37698	(37246)	s	**FC** FDCI	IM
37699	(37253)		**FC** FDCI	IM

Class 37/7. Refurbished locos. Main generator replaced by alternator. Regeared (CP7) bogies. Ballast weights added.

Details as class 37/4 except:
Main Alternator: GEC G564AZ (37796 – 803) Brush BA1005A (others).
Max. Tractive Effort: 276 kN (62000 lbf).
Weight: 120 t.   **RA:** 7.
All have twin fuel tanks.

37701	(37030)	s	**FC** LNCK	CF	
37702	(37020)	s	**FC** LNCK	CF	Taff Merthyr
37703	(37067)	s	**FC** ESBB	SL	
37704	(37034)	s	**FC** LNCK	CF	
37705	(37060)	a	**FP** ESBB	SL	
37706	(37016)	a	**FP** FDCI	IM	Conidae
37707	(37001)	a	**FP** FDCI	IM	
37708	(37089)	a	**FP** FDCI	IM	
37709	(37014)	a	**FP** ESBB	SL	
37710	(37044)		**FP** FDCI	IM	
37711	(37085)		**FM** FDCI	IM	
37712	(37102)		**FP** LGPM	ML	Teesside Steelmaster
37713	(37052)		**FM** FDCI	IM	British Steel Workington
37714	(37024)		**FM** LGPM	ML	
37715	(37021)		**FP** ESBB	SL	British Petroleum
37716	(37094)		**FM** FMCY	TE	British Steel Corby
37717	(37050)		**FP** FDCI	IM	Stainless Pioneer
37718	(37084)		**FM** FMCY	TE	Hartlepool Pipe Mill
37719	(37033)	a	**FP** FDCI	IM	
37796	(37105)	s	**FC** LNCK	CF	
37797	(37081)	s	**FC** LNCK	CF	
37798	(37006)	s	**FC** ESBB	SL	
37799	(37061)	s	**FC** LNCK	CF	Sir Dyfed/County of Dyfed
37800	(37143)	s	**FC** ESBB	SL	
37801	(37173)	s	**FC** LGPM	ML	
37802	(37163)	s	**FC** LNCK	CF	
37803	(37208)	s	**FC** ESBB	SL	
37883	(37176)		**FP** FDCI	IM	
37884	(37183)		**FP** FDCI	IM	Gartcosh
37885	(37177)		**FP** FDCI	IM	
37886	(37180)		**FM** FDCI	IM	
37887	(37120)	s	**FC** LNCK	CF	Caerphilly Castle/Castell Caerffili
37888	(37135)		**FP** FDYX	IM	Petrolea
37889	(37233)		**FC** LNCK	CF	
37890	(37168)	a	**FP** ESBB	SL	The Railway Observer

37891 (37166)	**FP** ESBB	SL	
37892 (37149)	**FP** ESBB	SL	Ripple Lane
37893 (37237)	**FP** LGPM	ML	
37894 (37124) s	**FC** LNCK	CF	
37895 (37283) s	**FC** LNCK	CF	
37896 (37231) s	**FC** LNCK	CF	
37897 (37155) s	**FC** LNCK	CF	
37898 (37186) s	**FC** LNCK	CF	Cwmbargoed DP
37899 (37161) s	**FC** LNCK	CF	County of West Glamorgan/
			Sir Gorllewin Morgannwg

Class 37/9. Refurbished Locos. Fitted with manufacturers prototype power units and ballast weights. Main generator replaced by alternator. Details as class 37/0 except:
Engine: Mirrlees MB275T of 1340 kW (1800 hp) at 1000 rpm (37901 – 4), Ruston RK270T of 1340 kW (1800 hp) at 900 rpm (37905 – 6).
Main Alternator: Brush BA1005A (GEC G564, 37905/6).
Max. Tractive Effort: 279 kN (62680 lbf).
Cont. Tractive Effort: 184 kN (41250 lbf) at 11.4 mph.
Weight: 120 t.   **RA:** 7.
All have twin fuel tanks.

37901 (37150)	**FM** LNHK	CF	Mirrlees Pioneer
37902 (37148)	**FM** LNHK	CF	
37903 (37249)	**FM** LNHK	CF	
37904 (37125)	**FM** LNHK	CF	
37905 (37136) s	**FM** LNHK	CF	Vulcan Enterprise
37906 (37206) s	**FM** LNHK	CF	

CLASS 43   HST POWER CAR   Bo – Bo

Built: 1976 – 82 by BREL Crewe Works. Formerly numbered as coaching stock but now classified as locomotives. Include luggage compartment.
Engine: Paxman Valenta 12RP200L of 1680 kW (2250 hp) at 1500 rpm. (Mirrlees MB190 of 1680 kW (2250 hp) m) (Paxman 12VP185 of 2010 kW (2700 hp) *).
Main Alternator: Brush BA1001B.
Traction Motors: Brush TMH68 – 46 or GEC G417AZ (43124 – 151/180). Frame mounted.
Max. Tractive Effort: 80 kN (17980 lbf).
Cont. Tractive Effort: 46 kN (10340 lbf) at 64.5 mph.
Power At Rail: 1320 kW (1770 hp).   **ETH:** Non standard 3-phase system.
Brake Force: **Length over Buffers:** 17.79 m.
Weight: 70 t.   **Wheel Diameter:** 1020 mm.
Max. Speed: 125 mph.   **RA:** 5.
Train Brakes: Air.
Multiple Working: With one other similar vehicle.
Communication Equipment: All equipped with driver – guard telephone and cab to shore radio-telephone.

§ Modified to be able to remotely control a class 91 locomotive and to be remotely controlled by a class 91 locomotive.

43002	I	IWRP	LA	Top of the Pops
43003	I	IWRP	PM	
43004	I	IWRP	LA	Swan Hunter
43005	I	IWRP	LA	
43006	I	ICCS	EC	
43007	I	ICCS	EC	
43008	I	ICCS	EC	
43009	I	IWRP	LA	
43010	I	IWRP	LA	
43011	I	IWRP	LA	Reader 125
43012	I	IWRP	LA	
43013 §	I	ICCS	EC	
43014 §	I	ICCS	EC	
43015	I	IWRP	LA	
43016	I	IWRP	PM	Gwyl Gerddi Cymru 1992
				Garden Festival Wales 1992
43017	I	IWRP	LA	
43018	I	IWRP	LA	
43019	I	IWRP	LA	Dinas Abertawe/City of Swansea
43020	I	IWRP	LA	John Grooms
43021	I	IWRP	LA	
43022	I	IWRP	LA	
43023	I	IWRP	LA	County of Cornwall
43024	I	IWRP	LA	
43025	I	IWRP	LA	Exeter
43026	I	IWRP	LA	City of Westminster
43027	I	IWRP	LA	Glorious Devon
43028	I	IWRP	LA	
43029	I	IWRP	LA	
43030	I	IWRP	PM	
43031	I	IWRP	PM	
43032	I	IWRP	PM	The Royal Regiment of Wales
43033	I	IWRP	PM	
43034	I	IWRP	PM	
43035	I	IWRP	PM	
43036	I	IWRP	PM	
43037	I	IWRP	PM	
43038	I	IECP	NL	National Railway Museum
				The First Ten Years 1975 – 1985
43039	I	IECP	NL	
43040	I	IWRP	PM	Granite City
43041	I	IWRP	LA	City of Discovery
43042	I	IWRP	LA	
43043	I	IMLP	NL	
43044	I	IMLP	NL	Borough of Kettering
43045	I	IMLP	NL	The Grammar School Doncaster AD 1350
43046	I	IMLP	NL	
43047	I	IMLP	NL	Rotherham Enterprise
43048	I	IMLP	NL	
43049	I	IMLP	NL	Neville Hill
43050	I	IMLP	NL	

43051	I	IMLP	NL	The Duke and Duchess of York
43052	I	IMLP	NL	City of Peterborough
43053	I	IMLP	NL	Leeds United
43054	I	IMLP	NL	
43055	I	IMLP	NL	Sheffield Star
43056	I	IMLP	NL	University of Bradford
43057	I	IMLP	NL	Bounds Green
43058	I	IMLP	NL	
43059	I	IMLP	NL	
43060	I	IMLP	NL	County of Leicestershire
43061	I	IMLP	NL	City of Lincoln
43062	I	ICCS	EC	
43063	I	ICCS	EC	
43064	I	IMLP	NL	City of York
43065 §	I	ICCS	EC	
43066	I	IMLP	NL	
43067 §	I	ICCS	EC	
43068 §	I	ICCS	EC	
43069	I	ICCS	EC	
43070	I	ICCS	EC	
43071	I	ICCS	EC	
43072	I	IMLP	NL	Derby Etches Park
43073	I	IMLP	NL	
43074	I	IMLP	NL	
43075	I	IMLP	NL	
43076	I	IMLP	NL	BBC East Midlands Today
43077	I	IMLP	NL	County of Nottingham
43078	I	ICCS	EC	Shildon County Durham
43079	I	ICCS	EC	
43080 §	I	ICCS	EC	
43081	I	IMLP	NL	
43082	I	IMLP	NL	
43083	I	IMLP	NL	
43084 §	I	ICCS	EC	County of Derbyshire
43085	I	IMLP	NL	City of Bradford
43086	I	ICCP	NL	
43087	I	ICCP	NL	
43088	I	ICCP	NL	XIII Commonwealth Games Scotland 1986
43089	I	ICCP	NL	
43090	I	ICCS	EC	
43091	I	ICCS	EC	Edinburgh Military Tattoo
43092	I	ICCS	EC	
43093	I	ICCS	EC	York Festival '88
43094	I	ICCS	EC	
43095	I	IECP	NL	
43096	I	IECP	NL	The Queens Own Hussars
43097	I	ICCS	EC	
43098	I	ICCS	EC	
43099	I	ICCS	EC	
43100	I	ICCS	EC	Craigentinny
43101	I	ICCP	NL	Edinburgh International Festival

43102	I	ICCP	NL	
43103	I	ICCP	NL	John Wesley
43104	I	IECP	NL	County of Cleveland
43105	I	IECP	NL	Hartlepool
43106	I	IECP	NL	Songs of Praise
43107	I	IECP	NL	
43108	I	IECP	NL	
43109	I	IECP	NL	Yorkshire Evening Press
43110	I	IECP	EC	Darlington
43111	I	IECP	EC	
43112	I	IECP	EC	
43113	I	IECP	EC	City of Newcastle-upon-Tyne
43114	I	IECP	EC	National Garden Festival Gateshead 1990
43115	I	IECP	EC	Yorkshire Cricket Academy
43116	I	IECP	EC	City of Kingston Upon Hull
43117	I	IECP	EC	
43118	I	IECP	EC	Charles Wesley
43119	I	IECP	EC	
43120	I	IECP	EC	
43121	I	ICCP	NL	West Yorkshire Metropolitan County
43122	I	ICCP	NL	South Yorkshire Metropolitan County
43123 §	I	ICCS	EC	
43124	I	IWRP	PM	
43125	I	IWRP	PM	City of Bristol
43126	I	IWRP	PM	City of Bristol
43127	I	IWRP	PM	
43128	I	IWRP	PM	
43129	I	IWRP	PM	
43130	I	IWRP	PM	Sulis Minerva
43131	I	IWRP	PM	Sir Felix Pole
43132	I	IWRP	PM	Worshipful Company of Carmen
43133	I	IWRP	PM	
43134	I	IWRP	PM	County of Somerset
43135	I	IWRP	PM	
43136	I	IWRP	PM	
43137	I	IWRP	PM	
43138	I	IWRP	PM	
43139	I	IWRP	PM	
43140	I	IWRP	PM	
43141	I	IWRP	PM	
43142	I	IWRP	PM	
43143	I	IWRP	PM	
43144	I	IWRP	PM	
43145	I	IWRP	PM	
43146	I	IWRP	PM	
43147	I	IWRP	PM	The Red Cross
43148	I	IWRP	PM	
43149	I	IWRP	PM	B.B.C. Wales Today
43150	I	IWRP	PM	Bristol Evening Post
43151	I	IWRP	PM	
43152	I	IWRP	PM	St. Peters School York AD 627

43153	I	ICCP	LA	University of Durham
43154	I	ICCP	LA	INTERCITY
43155	I	ICCP	NL	B.B.C. Look North
43156	I	ICCP	NL	
43157	I	ICCP	LA	Yorkshire Evening Post
43158	I	ICCP	LA	
43159	I	ICCP	LA	
43160	I	ICCP	LA	Storm Force
43161	I	ICCP	LA	Reading Evening Post
43162	I	ICCP	LA	Borough of Stevenage
43163	I	IWRP	LA	
43164	I	IWRP	LA	
43165	I	IWRP	LA	
43166	I	IWRP	LA	
43167 m	I	IXXA	PM (U)	
43168 m	I	IWRP	PM	
43169 m	I	IWRP	PM	The National Trust
43170 *	I	IWRP	LA	
43171	I	IWRP	LA	
43172	I	IWRP	LA	
43173	I	IWRP	LA	
43174	I	IWRP	LA	
43175	I	IWRP	LA	
43176	I	IWRP	LA	
43177	I	IWRP	LA	
43178	I	ICCP	LA	
43179	I	IWRP	LA	Pride of Laira
43180	I	ICCP	NL	
43181	I	IWRP	LA	Devonpart Royal Dockyard 1693-1993
43182	I	IWRP	LA	
43183	I	IWRP	LA	
43184	I	ICCP	LA	
43185	I	IWRP	LA	Great Western
43186	I	IWRP	LA	Sir Francis Drake
43187	I	IWRP	LA	
43188	I	IWRP	LA	City of Plymouth
43189	I	IWRP	LA	
43190	I	IWRP	LA	
43191	I	IWRP	LA	Seahawk
43192	I	IWRP	LA	City of Truro
43193	I	ICCP	LA	
43194	I	ICCP	LA	
43195	I	ICCP	LA	
43196	I	ICCP	LA	The Newspaper Society Founded 1836
43197	I	ICCP	LA	
43198	I	ICCP	NL	

CLASS 47   BRUSH TYPE 4   Co-Co

Built: 1963 – 67 by Brush Traction, Loughborough or BR Crewe Works.
Engine: Sulzer 12LDA28C of 1920 kW (2580 hp) at 750 rpm.

Main Generator: Brush TG160-60 Mk2, TG 160-60 Mk4 or TM172-50 Mk1.
Traction Motors: Brush TM64-68 Mk1 or Mk1A (axle hung).
Max. Tractive Effort: 267 kN (60000 lbf).
Cont. Tractive Effort: 133 kN (30000 lbf) at 26 mph.
Power At Rail: 1550 kW (2080 hp).   **Length over Buffers:** 19.38 m.
Brake Force: 61 t.   **Wheel Diameter:** 1143 mm.
Design Speed: 95 mph.   **Weight:** 120.5 – 125 t.
Max. Speed: various.   **RA:** 6 or 7.
Train Brakes: Air & vacuum.
Multiple Working: Not equipped (Blue Star Coupling Code*).
ETH Index (47/4, 47/6 and 47/7): 66 (75 Class 47/6).
Communication Equipment: Cab to shore radio-telephone.

Non standard liveries:

47145 is BR blue with black cab surrounds and Railfreight general markings (red and yellow bars).
47803 is grey, red and yellow.

Formerly numbered 1100 – 11, 1500 – 1999 not in order. 47299 was previously 47216. 47300 was previously 47468.

Class 47/0. Built with train heating boiler. RA6. Max Speed 75 mph.
a Vacuum brake isolated (Class 47/2).

47004	**G**	EWRS	SF	Old Oak Common Traction & Rolling Stock Depot
47016	**FO**	EWRS	SF	ATLAS
47033	a + **FD**	DAUT	TI	
47049	+ **FD**	DATT	TI	
47050	a + **FD**	DAYX	TI (U)	
47051	a + **FD**	DAUT	TI	
47052	**FD**	DAST	TI	
47053	a + **FE**	DACT	TI	Dollands Moor International
47060	a **FD**	DAST	TI	Halewood Silver Jubilee 1988
47063	**FA**	DAYX	TI (U)	
47079	**FD**	LBRB	BS	
47085	+ **FE**	DACT	TI	REPTA 1893 – 1993
47095	+ **FD**	DATT	TI	Southampton WRD Quality Approved
47096		DAYX	TI (U)	
47102		DAYX	TI (U)	
47114	a + **FD**	DATT	TI	
47121		EWRS	SF	Pochard
47125	+ **FE**	DACT	TI	
47142	**FR**	DART	TI (S)	
47144	a + **FD**	DATT	TI	
47145	**O**	DAST	TI	
47146	a	DAST	TI	
47147	**FD**	DAST	TI	
47156	a + **FD**	DATT	TI	
47157	**F**	DART	TI (S)	
47186	a + **FE**	DACT	TI	Catcliffe Demon
47187	**FD**	DAST	TI	

47188 a + FD	DATT	TI	
47190 FP	DAYX	TI (U)	
47193 FP	LBCB	BS	
47194 + FD	DATT	TI	Carlisle Currock Quality Approved
47197 FP	FDDI	IM	
47200 a + FD	DAUT	TI	
47201 + FD	DATT	TI	
47206 FD	DAST	TI	The Morris Dancer
47207 FD	LBRB	BS	
47210 + FD	DATT	TI	
47212 + FP	FDDI	IM	
47213 + FD	DATT	TI	Marchwood Military Port
47214 FD	DAYX	TI (U)	
47217 a + FE	DACT	TI	
47218 + FD	DATT	TI	United Transport Europe
47219 a + FD	DAUT	TI	Arnold Kunzler
47221 + FP	FDDI	IM	
47222 a + FD	DAUT	TI	
47223 + FD	EWAS	SF	
47224 + FP	FDDI	IM	
47225 FD	DAST	TI	
47226 a + FD	DATT	TI	
47228 + FD	DATT	TI	
47229 a + FD	DATT	TI	
47231 FD	DAST	TI	
47234 + FE	DACT	TI	
47236 a + FD	DAUT	TI	
47237 a + FD	DAUT	TI	
47238 FD	LBRB	BS	
47241 a + FE	DAUT	TI	
47245 + FE	DACT	TI	
47249 FR	DART	TI (S)	
47256 FD	FDRI	IM	
47258 + FD	DATT	TI	
47270	DART	TI (S)	
47276 + FP	FDRI	IM (U)	
47277 FD	FDDI	IM	
47278 FP	EWAS	SF	
47279 FD	DAST	TI	
47280 a + FD	DAUT	TI	Pedigree
47281 + FD	DATT	TI	
47283 FD	DAST	TI	Johnnie Walker
47284 + FD	DATT	TI	
47285 a + FD	DAUT	TI	
47286 a + FE	DACT	TI	Port of Liverpool
47287 + FD	DATT	TI	
47288 FD	DAYX	TI (U)	
47289 a FD	DAYX	TI (U)	
47290 a + FE	DACT	TI	
47291 a + FD	DATT	TI	The Port of Felixstowe
47292 a + FD	DATT	TI	

47293	a + **FD**	DATT	TI	
47294	+ **FD**	FDDI	IM	
47295	+ **FP**	LBCB	BS	
47296	**FD**	DAST	TI	
47297	a + **FD**	DATT	TI	
47298	+ **FD**	DATT	TI	Pegasus
47299	a + **FE**	DACT	TI	

Class 47/3. Built without Train Heat. (except 47300). RA6. Max Speed 75 mph. All equipped with slow speed control.

a Vacuum brake isolated (Class 47/2).

47300	**C**	LBRB	BS	
47301	**FR**	DART	TI	
47302	a **FR**	LBCB	BS	
47305	**FP**	DAST	TI	
47306	a + **FE**	DATT	TI	The Sapper
47307	a + **FE**	DACT	TI	
47308	**F**	LBCB	BS	
47310	a + **FD**	DAUT	TI	Henry Ford
47312	a + **FD**	DATT	TI	
47313	a + **FD**	DAUT	TI	
47315	**C**	EWAS	SF	Templecombe
47316	a + **FD**	DAUT	TI	
47317	**FD**	DART	TI (S)	Willesden Yard
47319	+ **FP**	FDEI	IM	Norsk Hydro
47321	**F**	DAYX	TI (U)	
47322	**FR**	DART	TI (S)	
47323	a + **FE**	DAUT	TI	ROVER GROUP QUALITY ASSURED
47325	**FO**	DAYX	TI (U)	
47326	a + **FD**	DAUT	TI	
47329	**C**	LBCB	BS	
47331	**C**	FDEI	IM	
47332	**C**	LBCB	BS	
47333	**C**	LBCB	BS	Civil Link
47334	**C**	LBCB	BS	
47335	a + **FD**	DATT	TI	
47337	**FO**	DAST	TI	
47338	a + **FE**	DAUT	TI	Warrington Yard
47339	**FD**	DAST	TI	
47340	**C**	LBCB	BS	
47341	**C**	LBCB	BS	
47344	a + **FE**	DACT	TI	
47345	**FR**	DART	TI (S)	
47346	**C**	FDEI	IM	
47347	a **FM**	DAST	TI	
47348	**FO**	EWRS	SF	St. Christopher's Railway Home
47349	**FD**	DAST	TI	
47350	**FO**	DART	TI (S)	
47351	a + **FE**	DACT	TI	
47352	**C**	FDRI	IM (U)	
47353	**C**	LBCB	BS	

47354 a	**FD**	DAST	TI	
47356	**FO**	LBRB	BS	
47357	**C**	LNXX	BS	The Permanent Way Institution
47358	**FO**	DART	TI	
47359	**FD**	FDRI	IM	
47360 a +	**FD**	DATT	TI	
47361 a +	**FD**	DATT	TI	Wilton Endeavour
47362 a +	**FD**	DAUT	TI	
47363 +	**F**	DATT	TI	
47365 +	**FE**	DACT	TI	ICI Diamond Jubilee
47366	**FO**	EWRS	SF	Capital Radio's
				Help a London Child
47367	**FR**	DAST	TI	
47368	**FP**	EWAS	SF	
47369	**FD**	FDEI	IM	
47370	**FO**	LBRB	BS	
47371	**FO**	DART	TI	
47372	**C**	LBCB	BS	
47375 a +	**FE**	DATT	TI	Tinsley Traction Depot
				Quality Approved
47376		DAYX	TI (U)	
47377 a	**FD**	DAST	TI	
47378 a +	**FD**	DATT	TI	
47379 +	**FP**	FDYX	IM (U)	

Class 47/0 and 47/3. Renumbered locos. RA6. Max Speed 75 mph. Fitted with blue star multiple working.
a Vacuum brake isolated (Class 47/2).

47387 (47314) a +	**FD**	DATT	TI	Transmark
47388 (47204) +	**FD**	DATT	TI	
47389 (47309) +	**FD**	DATT	TI	The Halewood Transmission
47390 (47330) a +	**FD**	DAMT	TI	Amlwch Freighter/Trên
				Nwyddau Amlwch (opp. sides)
47391 (47355) +	**FD**	DAMT	TI	
47392 (47304) a +	**FD**	DAMT	TI	Cory Brothers 1842 – 1992
47393 (47209) +	**FD**	DAMT	TI	Herbert Austin
47394 (47211) +	**FD**	DAMT	TI	Johnson Stevens Agencies
47395 (47205) +	**FD**	DAMT	TI	
47396 (47328) a +	**FD**	DAMT	TI	
47397 (47303) a +	**F**	DAMT	TI	
47398 (47152) a +	**FD**	DAMT	TI	
47399 (47150) a +	**FD**	DAMT	TI	

Class 47/4. Equipped with train heating. RA6. Max Speed 95 mph (§ 75 mph).

47462	**R**	EWAS	SF	
47463		PXLD	CD (U)	
47467	**BR**	PXLC	CD	
47471 §	**IO**	PXLH	CD	Norman Tunna G.C.
47473	**BR**	LBCB	BS	
47474	**R**	PXLC	CD	Sir Rowland Hill
47475	**RX**	PXLC	CD	Restive

Number		Code	Letters	Depot	Name
47476		**R**	PXLC	CD	Night Mail
47478			LBCB	BS	
47481	§	**BR**	PXLH	CD	
47484		**G**	EWRS	SF	ISAMBARD KINGDOM BRUNEL
47489		**R**	PXLC	CD	Crewe Diesel Depot
47492	§	**IO**	PXLH	CD	
47501	§	**R**	PXLH	CD	Craftsman
47513	§	**BR**	PXLH	CD	Severn
47519	§	**BR**	PXLH	CD	
47520		**I**	PXLC	CD	Thunderbird
47521		**RX**	PXLD	CD (U)	
47522	§	**R**	PXLH	CD	Doncaster Enterprise
47523		**M**	PXLC	CD	
47524		**RX**	PXLC	CD	
47525		**IO**	LBRB	BS	
47526		**BR**	EWRS	SF	
47528		**M**	PXLC	CD	The Queen's Own Mercian Yeomanry
47530		**RX**	PXLC	CD	
47532		**RX**	PXLC	CD	
47535		**R**	PXLD	CD (U)	University of Leicester
47536	§	**BR**	PXLH	CD	
47539	§	**RX**	PXLH	CD	
47543		**R**	PXLC	CD	
47547	§	**N**	PXLH	CD	
47550		**M**	FDRI	IM	University of Dundee
47555 (47126)		**IO**	FDYX	IM (U)	
47557 (47024)		**RX**	PXLC	CD	
47558 (47027)		**RX**	PXLC	CD	
47565 (47039)		**RX**	PXLC	CD	Responsive
47566 (47043)		**RX**	PXLC	CD	
47567 (47044)		**RX**	PXLC	CD	Red Star ISO 9002
47568 (47045)		**RX**	PXLC	CD	Royal Logistic Corps Postal & Courier Services
47572 (47168)		**R**	PXLC	CD	Ely Cathedral
47574 (47174)		**R**	PXLC	CD	Benjamin Gimbert G.C.
47575 (47175)		**R**	PXLC	CD	City of Hereford
47576 (47176)		**RX**	PXLC	CD	
47579 (47183)		**N**	EWTS	SF	James Nightall G.C.
47582 (47170)		**R**	PXLC	CD	County of Norfolk
47583 (47172)		**RX**	PXLC	CD	
47584 (47180)		**RX**	PXLC	CD	
47588 (47178)		**RX**	PXLC	CD	Resurgent
47594 (47035)		**RX**	PXLC	CD	Resourceful
47596 (47255)		**RX**	PXLC	CD	
47598 (47182)		**RX**	PXLC	CD	
47624 (47087)		**RX**	PXLC	CD	Saint Andrew
47625 (47076)		**RX**	PXLC	CD	Resplendent
47626 (47082)		**RX**	PXLC	CD	
47627 (47273)		**RX**	PXLC	CD	
47628 (47078)		**RX**	PXLC	CD	

47634 (47158)	**R**	PXLC	CD	Holbeck
47635 (47029)	**R**	PXLC	CD	
47640 (47244)	**R**	PXLC	CD	University of Strathclyde

Class 47/6. Fitted with high phosphorus brake blocks. RA6. 75 mph.

47671 (47616)	**BR**	PXLC	CD	
47673 (47593)	**I0**	PXLH	CD	York Intercity Control
47674 (47604)	**BR**	EWAS	SF	Women's Royal Voluntary Service
47675 (47595) + **M**	PXLB	CD	
47676 (47586)	**I**	FDDI	IM	Northamptonshire
47677 (47617)	**I**	FDDI	IM	University of Stirling

Class 47/7. Fitted with an older form of TDM. RA6. 95 mph (75 mph §).

47701 (47493)	+ **RX**	PXLB	CD	
47702 (47504)	+ **N**	EWTS	SF	County of Suffolk
47703 (47514)	+ **R**	PXLB	CD	The Queen Mother
47704 (47495)	+ **RX**	PXLB	CD	
47705 (47554)	+ **RX**	PXLB	CD	
47706 (47494)	+ **PS**	PXLD	CD (U)	
47707 (47506)	§ + **RX**	PXLH	CD	Holyrood
47708 (47516)	+ **N**	PXLD	CD (U)	
47709 (47499)	+ **RX**	PXLB	CD	
47710 (47496)	+ **W**	PXLB	CD	
47711 (47498)	+ **N**	EWTS	SF	County of Hertfordshire
47712 (47505)	+ **R**	PXLB	CD	Lady Diana Spencer
47714 (47511)	§ + **RX**	PXLH	CD	
47715 (47502)	+ **N**	PXLB	CD	Haymarket
47716 (47507)	§ + **RX**	PXLH	CD	
47717 (47497)	§ + **R**	PXLH	CD	

Class 47/7. Parcels dedicated locos. RA6. 95 mph.

47721 (47)				
47722 (47)				
47723 (47)				
47724 (47)				
47725 (47)				
47726 (47)				
47727 (47569)	a + **RX**	PXLB	CD	
47728 (47)				
47729 (47)				
47730 (47)				
47731 (47)				
47732 (47580)	+ **RX**	PXLB	CD	
47733 (47)				
47734 (47)				
47735 (47)				
47736 (47587)	a + **RX**	PXLB	CD	Cambridge Traction & Rolling Stock Depot
47737 (47)				

47738 (47592)	a + **RX**	PXLB	CD	Bristol Barton Hill
47739 (47)				
47740 (47)				
47741 (47597)	+ **RX**	PXLB	CD	Resilient
47742 (47)				
47743 (47599)	a + **RX**	PXLB	CD	
47744 (47600)	a + **RX**	PXLB	CD	
47745 (47603)	+ **RX**	PXLB	CD	
47746 (47605)	a + **RX**	PXLB	CD	
47747 (47615)	a + **RX**	PXLB	CD	
47748 (47)				
47749 (47)				
47750 (47)				
47751 (47)				
47752 (47)				
47753 (47)				
47754 (47)				
47755 (47)				
47756 (47644)	+ **RX**	PXLB	CD	
47757 (47585)	a + **RX**	PXLB	CD	Restitution
47758 (47517)	+ **RX**	PXLB	CD	
47759 (47559)	+ **RX**	PXLB	CD	
47760 (47562)	+ **RX**	PXLB	CD	Restless
47761 (47564)	+ **RX**	PXLB	CD	
47762 (47573)	+ **RX**	PXLB	CD	
47763 (47581)	+ **RX**	PXLB	CD	
47764 (47630)	+ **RX**	PXLB	CD	Resounding
47765 (47631)	+ **RX**	PXLB	CD	Ressalder
47766 (47642)	+ **RX**	PXLB	CD	Resolute
47767 (47641)	+ **RX**	PXLB	CD	
47768 (47490)	+ **RX**	PXLB	CD	Resonant
47769 (47491)	+ **RX**	PXLB	CD	Resolve
47770 (47500)	+ **RX**	PXLB	CD	Reserved
47771 (47503)	+ **RX**	PXLB	CD	Heaton Traincare Depot
47772 (47537)	+ **RX**	PXLB	CD	
47773 (47541)	+ **RX**	PXLB	CD	
47774 (47551)	+ **RX**	PXLB	CD	Poste Restante
47775 (47531)	+ **RX**	PXLB	CD	Respite
47776 (47578)	+ **RX**	PXLB	CD	Respected
47777 (47636)	+ **RX**	PXLB	CD	Restored
47778 (47606)	+ **RX**	PXLB	CD	Irresistible
47779 (47612)	+ **RX**	PXLB	CD	
47780 (47618)	+ **RX**	PXLB	CD	
47781 (47653)	+ **RX**	PXLB	CD	
47782 (47824)	+ **RX**	PXLB	CD	
47783 (47809)	+ **I**	PXLB	CD	Finsbury Park
47784 (47)				
47785 (47)				
47786 (47)				
47787 (47)				
47788 (47)				

```
47789 (47    )
47790 (47    )
47791 (47    )
```

Class 47/4 continued. RA6. 95 mph.

47802	(47552)	+ I	EWRS	SF
47803	(47553)	+ 0	EWRS	SF
47804	(47591)	+ I	EWRS	SF
47805	(47650)	+ I	ILRA	BR	Bristol Bath Road
47806	(47651)	+ I	ILRA	BR
47807	(47652)	a + I	ILRA	BR
47810	(47655)	+ I	ILRA	BR
47811	(47656)	+ I	ILRA	BR
47812	(47657)	+ I	ILRA	BR
47813	(47658)	a + I	ILRA	BR
47814	(47659)	+ I	ILRA	BR
47815	(47660)	a + I	ILRA	BR
47816	(47661)	+ I	ILRA	BR
47817	(47662)	+ I	ILRA	BR
47818	(47663)	+ I	ILRA	BR
47819	(47664)	+ I	PXLB	CD
47820	(47665)	a + I	PXLB	CD
47821	(47607)	a + I	PXLB	CD	Royal Worcester
47822	(47571)	+ I	ILRA	BR
47823	(47610)	+ I	PXLB	CD	SS Great Britain
47825	(47590)	a + I	ILRA	BR	Thomas Telford
47826	(47637)	a + I	ILRA	BR
47827	(47589)	+ I	ILRA	BR
47828	(47629)	a + I	ILRA	BR
47829	(47619)	a + I	ILRA	BR
47830	(47649)	+ I	ILRA	BR
47831	(47563)	+ I	ILRA	BR	Bolton Wanderer
47832	(47560)	a + I	ILRA	BR	Tamar
47833	(47608)	a + G	PXLB	CD	Captain Peter Manisty RN
47834	(47609)	+ I	PXLD	CD	FIRE FLY
47835	(47620)	a + I	PXLD	CD	Windsor Castle
47839	(47621)	+ I	ILRA	BR
47840	(47613)	+ I	ILRA	BR
47841	(47622)	a + I	ILRA	BR	The Institution of Mechanical Engineers
47843	(47623)	a + I	ILRA	BR
47844	(47556)	+ I	ILRA	BR	Derby & Derbyshire Chamber of Commerce & Industry
47845	(47638)	a + I	ILRA	BR	County of Kent
47846	(47647)	+ I	ILRA	BR	THOR
47847	(47577)	+ I	ILRA	BR
47848	(47632)	a + I	ILRA	BR
47849	(47570)	+ M	ILRA	BR
47850	(47648)	+ I	ILRA	BR
47851	(47639)	+ I	ILRA	BR
47853	(47614)	a + M	ILRA	BR

47971	(97480)	*	**BR** CDJC	CD	Robin Hood
47972	(97545)		**CS** CDJC	CD	The Royal Army Ordance Corps
47973	(97561)		**M** CDJC	CD	Derby Evening Telegraph
47975	(47540)	*	**C** CDJC	CD	The Institution of Civil Engineers
47976	(47546)	*	**C** CDJC	CD	Aviemore Centre

Class 47/3 continued. 100 mph loco for research purposes.

47981	(47364)	**C** CDJC	CD	

CLASS 56   BRUSH TYPE 5   Co–Co

Built: 1976 – 84 by Electroputere at Craiova, Romania (as sub contractors for Brush) or BREL at Doncaster or Crewe Works.
Engine: Ruston Paxman 16RK3CT of 2460 kW (3250 hp) at 900 rpm.
Main Alternator: Brush BA1101A.
Traction Motors: Brush TM73-62.
Max. Tractive Effort: 275 kN (61800 lbf).
Cont. Tractive Effort: 240 kN (53950 lbf) at 16.8 mph.
Power At Rail: 1790 kW (2400 hp). **Length over Buffers:** 19.36 m.
Brake Force: 60 t. **Wheel Diameter:** 1143 mm.
Design Speed: 80 mph. **Weight:** 125 t.
Max. Speed: 80 mph. **RA:** 7.
Train Brakes: Air.
Multiple Working: Red Diamond coupling code.
Communication Equipment: Cab to shore radio-telephone.
All equipped with slow speed control.

§ Derated to 1790 kW (2400 hp).
* Derated to 2060 kW (2800 hp).

56001	**FA**	LNXX	CF (U)	Whatley
56003	**F**	FDYX	IM (U)	
56004		LWBK	CF	
56005	**FC**	FDBK	IM	
56006	**FC**	FDBK	IM	
56007	**FC**	LWBK	CF	
56008		FDYX	IM (U)	
56009	**FC**	LWBK	CF	
56010		LWBK	CF	
56011	**F**	FDBK	IM	
56012	**FC**	FDYX	IM (U)	
56014	**FC**	FDYX	IM (U)	
56016	**FC**	LNXX	CF (U)	
56018	**FC**	LWBK	CF	
56019	**FR**	LWBK	CF	
56020		LNXX	CF (U)	
56021	**FC**	FDBK	IM	
56022		LWBK	CF	
56024	**FO**	FDYX	IM (U)	
56025	**FC**	LWBK	CF	

56026		FDYX	IM (U)	
56027	FC	FDYX	IM (U)	
56029	F	LWBK	CF	
56031	C	FDBK	IM	Merehead
56032	FM	LNBK	CF	Sir De Morgannwg/ County of South Glamorgan
56033	FA	LWBK	CF	
56034	FA	FMBY	TE	Castell Ogwr/Ogmore Castle
56035	FA	FDBI	IM	
56036	C	LWBK	CF	
56037	FA	LWBK	CF	Richard Trevithick
56038	FM	LNBK	CF	Western Mail
56039	FA	FMBY	TE	
56040	FM	LNBK	CF	Oystermouth
56041	FA	FDBK	IM	
56043	F	FDBK	IM	
56044	FM	LNBK	CF	Cardiff Canton Quality Assured
56045	FA	FMBY	TE	
56046	C	FDBK	IM	
56047	C	LWBK	CF	
56048	C	FDBI	IM	
56049	C	LWBK	CF	
56050	FA	FMBY	TE	
56051	FA	FDBI	IM	Isle of Grain
56052	F	LNBK	CF	
56053	FM	LNBK	CF	Sir Morgannwg Ganol/ County of Mid Glamorgan
56054	FM	LWBK	CF	British Steel Llanwern
56055	FA	FDBK	IM	
56056	FA	LWBK	CF	
56057	FA	LGAM	ML	
56058	FA	LGAM	ML	
56059	FA	LWBK	CF	
56060	FM	LNBK	CF	The Cardiff Rod Mill
56061	FM	FDBI	IM	
56062	F	FMBY	IM	Mountsorrel
56063	FA	FMBY	TE	Bardon Hill
56064	FM	LNBK	CF	
56065	FA	FMBY	TE	
56066	FC	LNXX	CF	
56067	FC	FDBK	IM	
56068	FC	FDBK	IM	
56069	§ FM	FMBY	TE	Thornaby TMD
56070	FA	LWBK	CF	
56071	F	LWBK	CF	
56072	F	LGAM	ML	
56073	FM	LNBK	CF	Tremorfa Steelworks
56074	FC	FDBK	IM	Kellingley Colliery
56075	F	FDBK	IM	West Yorkshire Enterprise
56076	FM	LNBK	CF	British Steel Trostre
56077	§ FC	FDBK	IM	Thorpe Marsh Power Station

THE PLATFORM 5
TRANSPORT BOOK CLUB

The Platform 5 Transport Book Club is a service which enables customers to order new Platform 5 titles before publication, at discounts of between 15% and 25% of the normal retail price. Unlike other book clubs, a small annual subscription fee of £1.50 is charged to cover the production and postage of a quarterly newsletter, but there is no obligation whatsoever to buy any books at any time. We like customers to buy our books because of their quality, not because of any obligation to a book club.

To illustrate the sort of discounts available, the following offer applied to Exeter-Newton Abbot - A Railway History, when first published in November 1993:

Cover Price: £25.00
Less book Club Member Discount £5.00
. . . . £20.00
Plus Contribution to Postage & Packing £2.00
BOOK CLUB MEMBER PRICE **£22.00**

Even after our 10% postage and packing charge, this still represents a saving of £3.00 on the cover price and more than covers the annual subscription.

Reduced prices are only available to Platform 5 Transport Book Club members, and are only applicable if subscriptions are received by a certain date prior to publication, to be advised.

There is no limit to the number of copies of each book that may be ordered through the Platform 5 Transport Book Club.

Books ordered through the Platform 5 Transport Book Club will be despatched as soon as possible after publication.

Customers should be aware that although most new Platform 5 titles will be available via this service, some low-priced titles will be excluded and in particular, the British Rail Pocket Books will not be available at a discounted price. These books may still be ordered through the club, at the normal retail price.

THE PLATFORM 5
TRANSPORT BOOK CLUB
MEMBERSHIP APPLICATION FORM

To enrol for one year's membership in the PLATFORM 5 TRANSPORT BOOK CLUB, please complete this form (or a photocopy) and send it with your cheque/postal order for £1.50 made payable to 'Platform 5 Publishing Limited' to:

The Platform 5 Transport Book Club, Wyvern House, Sark Road, SHEFFIELD, S2 4HG.

BLOCK CAPITALS PLEASE

Name: . . . .

Address: . . . .

. . . .

Post Code: . . . .

Telephone Number: . . . .

Please accept my application and enrol me as a member of the Platform 5 Transport Book Club.

As a member I will receive four issues of the club newsletter, each containing a number of new books at prices of at least 15% less than the published cover price (exclusive of postage and packing).

I understand I am note obliged to buy any of the books offered, and there is no limit to the number of copies of each book that may be ordered.

I enclose my cheque/postal order for £1.50 payable to 'Platform 5 Publishing Limited'.

Signed: . . . .

Date: . . . .

Office Use Only: . . . .

PLATFORM 5 PUBLISHING LTD
MAIL ORDER

NEW TITLES | | Price

Tram to Supertram **AUGUST**	4.95
BR Pocket Book No.1: Locomotives	1.95
Diesel & Electric Loco Register 3rd edition	7.95
Thomas The Privatised Tank Engine (Midland)	4.99
Engineers Series Wagon Fleet 970000-999999 (SCTP)	6.95
Docklands Official Handbook (Capital)	7.95
Railways South East - The Album (Capital)	9.95
Bus Review 9 (Bus Enthusiast)	5.95
Metrobus - The Company's First Ten Years (Capital)	7.95

Modern British Railway Titles

British Railways Locomotives & Coaching Stock 1994	7.50
BR Pocket Book No.2: Coaching Stock	1.85
BR Pocket Book No.3: DMUs & Channel Tunnel Stock	1.85
BR Pocket Book No.4: Electric Multiple Units	1.85
Preserved Locomotives of British Railways 8th edition	6.95
On-Track Plant on British Railways 4th edition	5.50
The Fifty 50s in Colour	5.95
British Rail Passenger Trains (Capital)	7.95
London Underground Rolling Stock 13th edition (Capital)	8.95
Underground Official Handbook (Capital)	5.95
Departmental Coaching Stock 5th edition (SCTP)	6.95
British Rail Wagon Fleet - B-Prefix Series (SCTP)	6.95
British Rail Internal Users (SCTP)	7.95
RIV Wagon Fleet (SCTP)	5.95
Miles & Chains Volume 2 - London Midland	1.40
Miles & Chains Volume 3 - Scottish	1.00
Miles & Chains Volume 5 - Southern	1.00
Rails In The Isle Of Man (Midland)	14.99

Overseas Railways

German Railways Locomotives & MUs 3rd edition	12.50
Swiss Railways/Chemins de fer Suisses	9.95
French Railways/Chemins de fer Francais 2nd edition	14.95
Steam on four Continents 3: Asia	7.95
TGV Handbook (Capital)	7.95
Paris Metro Handbook (Capital)	7.95
Irish Narrow Gauge - Pictorial History Part 1 (Midland)	15.99
Irish Narrow Gauge - Pictorial History Part 2 IMidland)	15.99
Midland & Great Western Railway of Ireland (Midland)	18.99
Irish Railways In Colour 1955-67 (Midland)	14.99

Historical Railway Titles

Exeter-Newton Abbot - A Railway History	25.00
6203 'Princess Margaret Rose'	19.95
Midland Railway Portrait	12.95
Steam Days on BR 1 - The Midland Line in Sheffield	4.95
Rails along the Sea Wall (Dawlish-Teignmouth Pictorial)	4.95
The Rolling Rivers	6.95
British Baltic Tanks	6.95
The Railways of Winchester	6.95
LNWR Branch Lines of West Leics & East Warwicks (Milepost)	7.95
Rails Through The Clay (Capital)	25.00
The 1938 Tube Stock (Capital)	9.95
Metropolitan Steam Locomotives (Capital)	9.95
The First Tube (Capital)	4.95
Going Green (Capital)	5.95

Political

The Battle for the Settle & Carlisle	6.95

Rambling

Rambles by Rail 1 - The Hope Valley Line 1.95
Rambles by Rail 2 - Liskeard-Looe 1.95
Rambles by Rail 4 - The New Forest 1.95

Light Rail Transit & Trams

Light Rail Review 1 (Reprint) 6.95
Light Rail Review 2 .... 7.50
Light Rail Review 3 .... 7.50
Light Rail Review 4 .... 7.50
Light Rail Review 5 .... 7.50
UK Light Rail Systems No.1: Manchester Metrolink (reprint) 9.95
Manx Electric 8.95
Blackpool & Fleetwood By Tram 7.50

Cars, Buses and Ships

Buses in Britain (Capital) 19.95
London Trolleybus Routes (Capital) 18.95
London Bus Handbook Part 1 (Capital) 8.95
London Bus Handbook Part 2 (Capital) 9.95

Maps and Track Diagrams (Quail Map Company)

British Rail Tack Diagrams 1 - Scotland & Isle of Man 5.00
British Rail Track Diagrams 4 - London Midland 6.95
London Railway Map 5.95
China Railway Atlas 4.00
China Railway Station List 1988 .... 4.00
New Zealand Railway & Tramway Atlas 3.50
Czech Republic & Slovakia Railway Map 1.70
Greece Railway Map 1.20
Poland Railway Map 2.00
New York Railway Map 1.70
Estonia Railway Map 1.20
Latvia & Lithuania Railway Map 2.00
Portugal Railway Map 2.00

European Railway Atlases (Ian Allan)

European Railway Atlas: Germany, Austria, Switzerland 9.95
European Railway Atlas: Spain, Portugal, Italy, Greece 10.99
European Railway Atlas: Scandinavia & Eastern Europe 10.99

Calendars

Modern Traction Calendar 1995 (Rail Photoprints) 3.95
Steam Traction Calendar 1995 (Rail Photoprints) 3.95

PVC Book Covers

A6 Pocket Book Covers in Blue, Red, Green or Grey 0.80
Locomotives & CS Covers in Blue, Red, Green 1.00
A5 Book Covers in Blue, Red, Green or Grey 1.40

Locomotives & Coaching Stock Back Numbers

1985	2.95	1989	4.95
1986	3.30	1990	5.95
1987	3.30	1991	6.60
1988	3.95	1992	7.00

British Railways Locomotives & Coaching Stock 1993 .... 7.25

All these publications are available from shops, bookstalls or direct from: Mail Order Department, Platform 5 Publishing Ltd, Wyvern House, Sark Road, SHEFFIELD, S2 4HG, ENGLAND. Telephone: (+ 44) 0114-255-2625 Fax: (+ 44) 0114-255-2471. For a full list of titles available by mail order, please send SAE to the above address.
When ordering please add postage & packing: 10% UK; 20% Europe; 30% Rest of World, minimum 30p. Payment may be made by sterling cheque, money order, Eurocheque or British postal order payable to 'Platform 5 Publishing Ltd', we also accept payment by credit card (Access, Visa, Eurocard, Mastercard). US Dollar Cheques can only be accepted if the equivalent of £3.00 is added to cover currency conversion.

56078	F	FDBK	IM	
56079	FC	LGAM	ML	
56080	F	FDBK	IM	Selby Coalfield
56081	F	FMBY	TE	
56082	FC	FDBK	IM	
56083 *	FC	FDBK	IM	
56084 *	FC	FDBI	IM	
56085	FC	FDBI	IM	
56086 *	FC	LWBK	CF	
56087	FM	FMBY	TE	
56088	FC	FDBI	IM	
56089	FC	FDBI	IM	Ferrybridge C Power Station
56090	FC	FDBI	IM	
56091	F	FDBK	IM	Castle Donington Power Station
56092	F	LWBK	CF	
56093	F	LWBK	CF	The Institution of Mining Engineers
56094	FC	FDBI	IM	Eggborough Power Station
56095	F	FDBK	IM	Harworth Colliery
56096	FC	LGAM	ML	
56097	FM	FMBY	TE	
56098	FC	FDBK	IM	
56099	FC	LWBK	CF	Fiddlers Ferry Power Station
56100	FC	FDBK	IM	
56101	FC	LGAM	ML	Mutual Improvement
56102	F	FDBK	IM	Scunthorpe Steel Centenary
56103	FA	LGAM	ML	
56104	FC	LGAM	ML	
56105	FA	LWBK	CF	
56106	FC	FDBI	IM	
56107 §	FC	FMBY	TE	
56108	F	FMBB	IM	
56109	FC	FMBB	IM	
56110	FA	FMBB	IM	Croft
56111	FC	FMBB	IM	
56112	FC	FMBB	IM	
56113	FC	LNBK	CF	
56114	FC	LNBK	CF	Maltby Colliery
56115	FC	LNBK	CF	
56116	FC	FMBY	TE	
56117	FC	FMBB	IM	Wilton-Coalpower
56118	FC	FMBB	IM	
56119	FC	LNBK	CF	
56120	FC	FMBB	IM	
56121	FC	LGAM	ML	
56123	FC	LGAM	ML	Drax Power Station
56124	FC	LGAM	ML	
56125	FC	LWBK	CF	
56126	FC	FDBI	IM	
56127	F	LWBK	CF	
56128	FC	LGAM	ML	West Burton Power Station
56129	FC	LGAM	ML	

56130	FC	FMBB	IM	Wardley Opencast
56131	F	FMBB	IM	Ellington Colliery
56132	FC	LWBK	CF	
56133	F	LWBK	CF	Crewe Locomotive Works
56134	FC	FMBB	IM	Blyth Power
56135	F	FMBB	IM	Port of Tyne Authority

CLASS 58   BREL TYPE 5   Co – Co

Built: 1983 – 87 by BREL at Doncaster Works.
Engine: Ruston Paxman RK3ACT of 2460 kW (3300 hp) at 1000 rpm.
Main Alternator: Brush BA1101B.
Traction Motors: Brush TM73-62.
Max. Tractive Effort: 275 kN (61800 lbf).
Cont. Tractive Effort: 240 kN (53950 lbf) at 17.4 mph.
Power At Rail: 1780 kW (2387 hp). **Length over Buffers:** 19.13 m.
Brake Force: 62 t. **Wheel Diameter:** 1120 mm.
Design Speed: 80 mph. **Weight:** 130 t.
Max. Speed: 80 mph. **RA:** 7.
Train Brakes: Air.
Multiple Working: Red Diamond coupling code.
Communication Equipment: Cab to shore radio-telephone.
All equipped with slow speed control.

58001	FC	ENBN	TO	
58002	F	ENBN	TO	Daw Mill Colliery
58003	F	ENBN	TO	Markham Colliery
58004	FC	ENBN	TO	
58005	F	ENBN	TO	
58006	FC	ENBN	TO	
58007	FC	ENBN	TO	Drakelow Power Station
58008	FC	ENBN	TO	
58009	FC	ENBN	TO	
58010	FC	ENBN	TO	
58011	F	ENBN	TO	Worksop Depot
58012	F	ENBN	TO	
58013	FC	ENBN	TO	
58014	F	ENBN	TO	Didcot Power Station
58015	F	ENBN	TO	
58016	FC	ENBN	TO	
58017	FC	ENBN	TO	
58018	FC	ENBN	TO	High Marnham Power Station
58019	F	ENBN	TO	Shirebrook Colliery
58020	FC	ENBN	TO	Doncaster Works
58021	FC	ENBN	TO	
58022	F	ENBN	TO	
58023	F	ENBN	TO	
58024	FC	ENBN	TO	

58025	F	ENBN	TO	
58026	F	ENBN	TO	
58027	FC	ENBN	TO	
58028	FC	ENBN	TO	
58029	FC	ENBN	TO	
58030	F	ENBN	TO	
58031	F	ENBN	TO	
58032	F	ENBN	TO	
58033	FC	ENBN	TO	
58034	FC	ENBN	TO	Bassetlaw
58035	F	ENBN	TO	
58036	FC	ENBN	TO	
58037	FC	ENBN	TO	
58038	FC	ENBN	TO	
58039	FC	ENBN	TO	Rugeley Power Station
58040	F	ENBN	TO	Cottam Power Station
58041	F	ENBN	TO	Ratcliffe Power Station
58042	FC	ENBN	TO	Ironbridge Power Station
58043	FC	ENBN	TO	Knottingley
58044	FC	ENBN	TO	Oxcroft Opencast
58045	FC	ENBN	TO	
58046	FC	ENBN	TO	Thoresby Colliery
58047	FC	ENBN	TO	Manton Colliery
58048	FC	ENBN	TO	Coventry Colliery
58049	FC	ENBN	TO	Littleton Colliery
58050	FC	ENBN	TO	Toton Traction Depot

CLASS 59   GENERAL MOTORS TYPE 5   Co–Co

Built: 1985 (59001 – 4), 1989 (59005) by General Motors, La Grange, Illinois, U.S.A. or 1990 (59101 – 4) and 1994 (59201) by General Motors, London, Ontario, Canada.
Engine: General Motors 645E3C two stroke of 2460 kW (3300 hp) at 900 rpm.
Main Alternator: General Motors AR11 MLD-D14A.
Traction Motors: General Motors D77B.
Max. Tractive Effort: 506 kN (113 550 lbf).
Cont. Tractive Effort: 291 kN (65 300 lbf) at 14.3 mph.
Power At Rail: 1889 kW (2533 hp).   **Length over Buffers:** 21.35 m.
Brake Force: 69 t.   **Wheel Diameter:** 1067 mm.
Design Speed: 60 mph.   **Weight:** 121 t.
Max. Speed: 60 mph.   **RA:** 7.

Class 59/0. Owned by Foster-Yeoman Ltd. Blue/silver/blue livery with white lettering and cast numberplates.

59001	0	XYPO	MD	YEOMAN ENDEAVOUR
59002	0	XYPO	MD	YEOMAN ENTERPRISE
59003	0	XYPO	MD	YEOMAN HIGHLANDER
59004	0	XYPO	MD	YEOMAN CHALLENGER
59005	0	XYPO	MD	KENNETH J. PAINTER

Class 59/1. Owned by ARC Limited. Yellow/grey with grey lettering and cast numberplates.

59101	**0**	XYPA	WH	Village of Whatley
59102	**0**	XYPA	WH	Village of Chantry
59103	**0**	XYPA	WH	Village of Mells
59104	**0**	XYPA	WH	Village of Great Elm

Class 59/2. Owned by National Power. Grey, red, white and blue with white and red lettering and cast numberplates.

59201	**0**	XYPN	FB	Vale of York

CLASS 60   BRUSH TYPE 5   Co – Co

Built: 1989 onwards by Brush Traction.
Engine: Mirrlees MB275T of 2310 kW (3100 hp) at 1000 rpm.
Main Alternator: Brush.
Traction Motors: Brush separately excited.
Max. Tractive Effort: 500 kN (106500 lbf).
Cont. Tractive Effort: 336 kN (71570 lbf) at 17.4 mph.
Power At Rail: 1800 kW (2415 hp). **Length over Buffers:** 21.34 m.
Brake Force: 74 t. **Wheel Diameter:** 1118 mm.
Design Speed: 62 mph. **Weight:** 129 t.
Max. Speed: 60 mph. **RA:** 7.
Multiple Working: Within class.
Communication Equipment: Cab to shore radio-telephone.
All equipped with slow speed control.

60001	**FA**	ESAB	SL	Steadfast
60002	**FP**	FDAK	IM	Capability Brown
60003	**FP**	FDAI	IM	Christopher Wren
60004	**FC**	FDAI	IM	Lochnagar
60005	**FA**	LWCK	CF	Skiddaw
60006	**FA**	ENAN	TO	Great Gable
60007	**FP**	FMAY	TE	Robert Adam
60008	**FM**	FDAI	IM	Moel Fammau
60009	**FA**	ENAN	TO	Carnedd Dafydd
60010	**FA**	ENAN	TO	Pumlumon Plynlimon
60011	**FA**	ENAN	TO	Cader Idris
60012	**FA**	ENAN	TO	Glyder Fawr
60013	**FP**	FDAI	IM	Robert Boyle
60014	**FP**	FDAI	IM	Alexander Fleming
60015	**FA**	LWCK	CF	Bow Fell
60016	**FA**	LWCK	CF	Langdale Pikes
60017	**FA**	ENAN	TO	Arenig Fawr
60018	**FA**	ESAB	SL	Moel Siabod
60019	**FA**	ESAB	SL	Wild Boar Fell
60020	**FM**	FMAY	TE	Great Whernside
60021	**FM**	FDAI	IM	Pen-y-Ghent
60022	**FM**	FMAY	TE	Ingleborough
60023	**FM**	FMAY	TE	The Cheviot

60024	FP	FDAI	IM	Elizabeth Fry
60025	FP	FDAI	IM	Joseph Lister
60026	FP	FDAI	IM	William Caxton
60027	FP	FDAK	IM	Joseph Banks
60028	FP	FDAI	IM	John Flamstead
60029	FM	LNAK	CF	Ben Nevis
60030	FM	FMAY	TE	Cir Mhor
60031	FM	FMAY	TE	Ben Lui
60032	FC	LWAK	CF	William Booth
60033	FP	LNAK	CF	Anthony Ashley Cooper
60034	FM	LNAK	CF	Carnedd Llewelyn
60035	FM	LNAK	CF	Florence Nightingale
60036	FM	LNAK	CF	Sgurr Na Ciche
60037	FM	LNAK	CF	Helvellyn
60038	FM	FMAY	TE	Bidean Nam Bian
60039	FA	ESAB	SL	Glastonbury Tor
60040	FA	ESAB	SL	Brecon Beacons
60041	FA	ESAB	SL	High Willhays
60042	FA	ESAB	SL	Dunkery Beacon
60043	FA	ESAB	SL	Yes Tor
60044	FM	ENAN	TO	Ailsa Craig
60045	FC	LWAK	CF	Josephine Butler
60046	FC	LWAK	CF	William Wilberforce
60047	FC	LWAK	CF	Robert Owen
60048	FA	ENAN	TO	Saddleback
60049	FM	FMAY	TE	Scafell
60050	F	FDAI	IM	Roseberry Topping
60051	FP	FDAI	IM	Mary Somerville
60052	FM	FMAY	TE	Goat Fell
60053	FP	FDAI	IM	John Reith
60054	FP	FDAI	IM	Charles Babbage
60055	FC	LWAK	CF	Thomas Barnardo
60056	F	LWAK	CF	William Beveridge
60057	FC	LWAK	CF	Adam Smith
60058	FC	LWAK	CF	John Howard
60059	FC	FDAK	IM	Samuel Plimsoll
60060	FC	LWCK	CF	James Watt
60061	FC	LWAK	CF	Alexander Graham Bell
60062	FP	LNAK	CF	Samuel Johnson
60063	FP	LNAK	CF	James Murray
60064	FP	FDAK	IM	Back Tor
60065	FP	LNAK	CF	Kinder Low
60066	FC	LWAK	CF	John Logie Baird
60067	F	FDAK	IM	James Clerk-Maxwell
60068	F	FDAK	IM	Charles Darwin
60069	F	FDAK	IM	Humphry Davy
60070	FC	FDAK	IM	John Loudon McAdam
60071	F	ENAN	TO	Dorothy Garrod
60072	FC	ENAN	TO	Cairn Toul
60073	FC	ENAN	TO	Cairn Gorm
60074	FC	ENAN	TO	Braeriach

60075	FC	ENAN	TO	Liathach
60076	FC	ENAN	TO	Suilven
60077	FC	ENAN	TO	Canisp
60078	FC	ENAN	TO	Stac Pollaidh
60079	F	ENAN	TO	Foinaven
60080	FA	LWCK	CF	Kinder Scout
60081	FA	LNAK	CF	Bleaklow Hill
60082	FA	LWCK	CF	Mam Tor
60083	FA	ENAN	TO	Shining Tor
60084	FA	LWCK	CF	Cross Fell
60085	FA	LWCK	CF	Axe Edge
60086	FC	ENAN	TO	Schiehallion
60087	FC	ENAN	TO	Slioch
60088	FC	ENAN	TO	Buachaille Etive Mor
60089	FC	LWAK	CF	Arcuil
60090	FC	FMAY	TE	Quinag
60091	FC	FDAI	IM	An Teallach
60092	FC	LNAK	CF	Reginald Munns
60093	FC	LNAK	CF	Jack Stirk
60094	FA	ENAN	TO	Tryfan
60095	FA	LWCK	CF	Crib Goch
60096	FA	LNAK	CF	Ben Macdui
60097	FA	LWCK	CF	Pillar
60098	FA	ENAN	TO	Charles Francis Brush
60099	FA	ESAB	SL	Ben More Assynt
60100	FA	ESAB	SL	Boar of Badenoch

BR ELECTRIC LOCOMOTIVES

CLASS 73/0   ELECTRO – DIESEL   Bo – Bo

Built: 1962 by BR at Eastleigh Works.
Supply System: 660 – 850 V d.c. from third rail.
Engine: English Electric 4SRKT of 447 kW (600 hp) at 850 rpm.
Main Generator: English Electric 824/3D.
Traction Motors: English Electric 542A.
Max. Tractive Effort: Electric 187 kN (42000 lbf). Diesel 152 kN (34100 lbf).
Continuous Rating: Electric 1060 kW (1420 hp) giving a tractive effort of 43 kN (9600 lbf) at 55.5 mph.
Cont. Tractive Effort: Diesel 72 kN (16100 lbf) at 10 mph.
Maximum Rail Power: Electric 1830 kW (2450 hp) at 37 mph.

Brake Force: 31 t.	**Length over Buffers:** 16.36 m.
Design Speed: 80 mph.	**Weight:** 76.5 t.
Max. Speed: 60 mph.	**RA:** 6.
Wheel Diameter: 1016 mm.	**ETH Index (Elec. power):** 66

Train Brakes: Air, Vacuum and electro-pneumatic.
Multiple Working: Within sub-class, with Class 33/1 and various SR EMUs.
Communication Equipment: All equipped with driver – guard telephone.
Couplings: Drop-head buckeye.
Formerly numbered E 6001 – 3/5/6.

Non-standard Livery: 73005 is Network SouthEast blue.

73001	**MD**	HEBD	BD	
73002	**BR**	HEBD	BD	
73003	**G**	ENZX	SL	Sir Herbert Walker
73005	**O**	HEBD	BD	
73006	**MD**	HEBD	BD	

CLASS 73/1 & 73/2   ELECTRO – DIESEL   Bo – Bo

Built: 1965 – 67 by English Electric Co. at Vulcan Foundry, Newton le Willows.
Supply System: 660 – 850 V d.c. from third rail.
Engine: English Electric 4SRKT of 447 kW (600 hp) at 850 rpm.
Main Generator: English Electric 824/5D.
Traction Motors: English Electric 546/1B.
Max. Tractive Effort: Electric 179 kN (40000 lbf). Diesel 160 kN (36000 lbf).
Continuous Rating: Electric 1060 kW (1420 hp) giving a tractive effort of 35 kN (7800 lbf) at 68 mph.
Cont. Tractive Effort: Diesel 60 kN (13600 lbf) at 11.5 mph.
Maximum Rail Power: Electric 2350 kW (3150 hp) at 42 mph.

Brake Force: 31 t.	**Length over Buffers:** 16.36 m.
Design Speed: 90 mph.	**Weight:** 77 t.
Max. Speed: 60 (90*) mph.	**RA:** 6.
Wheel Diameter: 1016 mm.	**ETH Index (Elec. power):** 66

Train Brakes: Air, Vacuum and electro-pneumatic.
Multiple Working: Within sub-class, with Class 33/1 and various SR EMUs.

Communication Equipment: All equipped with driver – guard telephone.
Couplings: Drop-head buckeye.
Non-standard Livery: 73101 is Pullman umber & Cream.

Class 73/2 are locos dedicated to Gatwick Express services.

a Vacuum brake isolated.

Formerly numbered E 6001 – 20/22 – 26/28 – 49 (not in order).

73101	**0**	EWHB	SL	The Royal Alex'
73103	**I0**	EWEB	SL	
73104	**I0**	EWEB	SL	
73105	**C**	EWEB	SL	
73106	**D**	EWEB	SL	
73107	**C**	EWHB	SL	Redhill 1844 – 1994
73108	**C**	EWEB	SL	
73109	* **N**	HYSB	BM	Battle of Britain 50th Anniversary
73110	**C**	EWEB	SL	
73112	**N**	IVGA	SL (U)	University of Kent at Canterbury
73114	**I0**	EWEB	SL	
73117	**I0**	EWEB	SL	University of Surrey
73118	**C**	EWHB	SL	The Romney Hythe and Dymchurch Railway
73119	**C**	EWEB	SL	Kentish Mercury
73126	**N**	EWRB	SL	Kent & East Sussex Railway
73128	**C**	EWRB	SL	OVS BULLIED C.B.E. 1937 1949 C.M.E. SOUTHERN RAILWAY
73129	**N**	EWHB	SL	City of Winchester
73130	**C**	EWEB	SL	City of Portsmouth
73131	**C**	EWRB	SL	
73132	**I0**	EWRB	SL	
73133	**N**	EWEB	SL	The Bluebell Railway
73134	**I0**	EWEB	SL	Woking Homes 1885 – 1985
73136	**N**	EWEB	SL	Kent Youth Music
73138	**C**	EWEB	SL	
73139	**I0**	EWRB	SL	
73140	**I0**	EWRB	SL	
73141	**I0**	EWRB	SL	
73201 (73142)	a* **I**	IVGA	SL	Broadlands
73202 (73137)	a* **GE**	IVGA	SL	Royal Observer Corps
73203 (73127)	a* **I**	IVGA	SL	
73204 (73125)	a* **GE**	IVGA	SL	Stewarts Lane 1860 – 1985
73205 (73124)	* **M**	IVGA	SL	London Chamber of Commerce
73206 (73123)	a* **GE**	IVGA	SL	Gatwick Express
73207 (73122)	a* **GE**	IVGA	SL	County of East Sussex
73208 (73121)	a* **GE**	IVGA	SL	Croydon 1883 – 1983
73209 (73120)	a* **GE**	IVGA	SL	
73210 (73116)	a* **GE**	IVGA	SL	Selhurst
73211 (73113)	a* **I**	IVGA	SL	
73212 (73102)	a* **GE**	IVGA	SL	Airtour Suisse
73235 (73135)	a* **GE**	IVGA	SL	

NOTES FOR CLASSES 86 – 91.

The following common features apply to all locos of Classes 86 – 91.
Supply System: 25 kV a.c. from overhead equipment.
Communication Equipment: Driver – guard telephone and cab to shore radio-telephone.
Multiple Working: Time division multiplex system.
a vacuum brakes isolated.

CLASS 86/1   BR DESIGN   Bo – Bo

Built: 1965 – 66 by English Electric Co. at Vulcan Foundry, Newton le Willows or BR at Doncaster Works. Rebuilt with Class 87 type bogies and motors. Tap changer control.
Traction Motors: GEC G412AZ frame mounted.
Max. Tractive Effort: 258 kN (58000 lbf).
Continuous Rating: 3730 kW (5000 hp) giving a tractive effort of 95 kN (21300 lbf) at 87 mph.
Maximum Rail Power: 5860 kW (7860 hp) at 50.8 mph.

Brake Force: 40 t.	**Length over Buffers:** 17.83 m.
Design Speed: 110 mph.	**Weight:** 87 t.
Max. Speed: 110 mph.	**RA:** 6.
ETH Index: 74	**Wheel Diameter:** 1150 mm.
Train Brakes: Air & Vacuum.	**Electric Brake:** Rheostatic.

Note: Class 86 were formerly numbered E 3101 – 3200 (not in order).

86101 (86201)	I	IWPA	WN	Sir William A Stanier FRS
86102 (86202)	a I0	IWPA	WN	Robert A Riddles
86103 (86203)	I	IWPA	WN	André Chapelon

CLASS 86/2   BR DESIGN   Bo – Bo

Built: 1965 – 66 by English Electric Co. at Vulcan Foundry, Newton le Willows or BR at Doncaster Works. Later rebuilt with resilient wheels and flexicoil suspension. Tap changer control.
Traction Motors: AEI 282BZ.
Max. Tractive Effort: 207 kN (46500 lbf).
Continuous Rating: 3010 kW (4040 hp) giving a tractive effort of 85 kN (19200 lbf) at 77.5 mph.
Maximum Rail Power: 4550 kW (6100 hp) at 49.5 mph.

Brake Force: 40 t.	**Length over Buffers:** 17.83 m.
Design Speed: 125 mph.	**Weight:** 85 t – 86 t.
Max. Speed: 100 (110§) mph.	**RA:** 6.
ETH Index: 74	**Wheel Diameter:** 1156 mm.
Train Brakes: Air & Vacuum.	**Electric Brake:** Rheostatic.

86204	I	IWPA	WN	City of Carlisle
86205 (86503)	I	ICCA	LG	City of Lancaster
86206	I	ICCA	LG	City of Stoke on Trent
86207	a I	IWPA	WN	City of Lichfield

86208		a	I	PXLE	CE	City of Chester
86209		a§	M	IWPA	WN	City of Coventry
86210			I	PXLE	CE	City of Edinburgh
86212			I	ICCA	LG	Preston Guild 1328 – 1992
86213			I	IWPA	WN	Lancashire Witch
86214			I	ICCA	LG	Sans Pareil
86215			I	IANA	NC	Joseph Chamberlain
86216			I	ICCA	LG	Meteor
86217	(86504)		I	IANA	NC	City University
86218			I	IANA	NC	Harold MacMillan
86219		a	I	IWPA	WN	Phoenix
86220			I	IANA	NC	The Round Tabler
86221			I	IANA	NC	B.B.C. Look East
86222	(86502)		I	ICCA	LG	LLOYD'S LIST
						250th ANNIVERSARY
86223			I	IANA	NC	Norwich Union
86224		a§	I	IWPA	WN	Caledonian
86225		a§	I	IWPA	WN	Hardwicke
86226			M	ICCA	LG	Royal Mail Midlands
86227			M	ICCA	LG	Sir Henry Johnson
86228			I	ICCA	LG	Vulcan Heritage
86229			I	ICCA	LG	Sir John Betjeman
86230			I	IANA	NC	The Duke of Wellington
86231		a§	I	IWPA	WN	Starlight Express
86232			I	IANA	NC	Norwich Festival
86233	(86506)		I	ICCA	LG	Laurence Olivier
86234			I	ICCA	LG	J B Priestley OM
86235			I	IANA	NC	Crown Point
86236			I	IWPA	WN	Josiah Wedgwood
						MASTER POTTER 1736 – 1795
86237			I	IANA	NC	University of East Anglia
86238			I	IANA	NC	European Community
86239	(86507)		R	PXLE	CE	L S Lowry
86240		a	I	IWPA	WN	Bishop Eric Treacy
86241	(86508)		RX	PXLE	CE	Glenfiddich
86242			I	IWPA	WN	James Kennedy GC
86243			RX	PXLE	CE	
86244			I	ICCA	LG	The Royal British Legion
86245			I	IWPA	WN	Dudley Castle
86246	(86505)		I	IANA	NC	Royal Anglian Regiment
86247			I	ICCA	LG	Abraham Darby
86248			I	IWPA	WN	Sir Clwyd/County of Clwyd
86249		a	M	IWPA	WN	County of Merseyside
86250			I	IANA	NC	The Glasgow Herald
86251			I	IWPA	WN	The Birmingham Post
86252			I	ICCA	LG	The Liverpool Daily Post
86253	(86044)	a	I	IWPA	WN	The Manchester Guardian
86254	(86047)		RX	PXLE	CE	
86255	(86042)		I	ICCA	LG	Penrith Beacon
86256	(86040)		I	IWPA	WN	Pebble Mill
86257	(86043)		I	IWPA	WN	Snowdon

86258 (86501)	a l	IWPA	WN	Talyllyn – The First Preserved Railway
86259 (86045)	I	ICCA	LG	Peter Pan
86260 (86048)	I	ICCA	LG	Driver Wallace Oakes G.C.
86261 (86041)	**RX** PXLE		CE	

CLASS 86/4 & 86/6   BR DESIGN   Bo – Bo

Built: 1965 – 66 by English Electric Co. at Vulcan Foundry, Newton le Willows or BR at Doncaster Works. Later rebuilt with resilient wheels and flexicoil suspension. Tap changer control.
Traction Motors: AEI 282AZ.
Max. Tractive Effort: 258 kN (58000 lbf).
Continuous Rating: 2680 kW (3600 hp) giving a tractive effort of 89 kN (20000 lbf) at 67 mph.
Maximum Rail Power: 4400 kW (5900 hp) at 38 mph.

Brake Force: 40 t.	**Length over Buffers:** 17.83 m.
Design Speed: 100 mph.	**Weight:** 83 t – 84 t.
Max. Speed: 100 (75*) mph.	**RA:** 6.
ETH Index: 74	**Wheel Diameter:** 1156 mm.
Train Brakes: Air & Vacuum.	**Electric Brake:** Rheostatic.

Class 86/6 have the ETH equipment isolated.

86401 (86001)	**RX** PXLE	CE	
86602 (86402)	* **FD** DANC	CE	
86603 (86403)	* **FD** DANC	CE	
86604 (86404)	* **FD** DANC	CE	
86605 (86405)	* **FD** DANC	CE	Intercontainer
86606 (86406)	* **FD** DANC	CE	
86607 (86407)	* **FD** DANC	' CE	The Institution of Electrical Engineers
86608 (86408)	* **FE** DANC	CE	St. John Ambulance
86609 (86409)	* **FD** DANC	CE	
86610 (86410)	* **FD** DANC	CE	
86611 (86411)	* **FD** DANC	CE	Airey Neave
86612 (86412)	* **FD** DANC	CE	Elizabeth Garrett Anderson
86613 (86413)	* **FD** DANC	CE	County of Lancashire
86614 (86414)	* **FD** DANC	CE	Frank Hornby
86615 (86415)	* **FD** DANC	CE	Rotary International
86616 (86316)	**RX** PXLE	CE	
86417 (86317)	**RX** PXLE	CE	
86618 (86418)	* **FD** DANC	CE	
86619 (86319)	**RX** PXLE	CE	
86620 (86420)	* **FD** DANC	CE	
86621 (86421)	* **FD** DANC	CE	London School of Economics
86622 (86422)	* **FD** DANC	CE	
86623 (86423)	* **FD** DANC	CE	
86424 (86324)	**RX** PXLE	CE	
86425 (86325)	**RX** PXLE	CE	
86426 (86326)	**RX** PXLE	CE	
86627 (86427)	* **FD** DANC	CE	The Industrial Society

86628	(86428)	* **FD**	DANC	CE	Aldaniti
86430	(86030)	**RX**	PXLE	CE	
86631	(86431)	* **FE**	DANC	CE	
86632	(86432)	* **FD**	DANC	CE	Brookside
86633	(86433)	* **FD**	DANC	CE	Wulfruna
86634	(86434)	* **FD**	DANC	CE	University of London
86635	(86435)	* **FD**	DANC	CE	
86636	(86436)	* **FD**	DANC	CE	
86637	(86437)	* **FD**	DANC	CE	
86638	(86438)	* **FD**	DANC	CE	
86639	(86439)	* **FD**	DANC	CE	

CLASS 87   BR DESIGN   Bo–Bo

Built: 1973 – 75 by BREL at Crewe Works. Class 87/1 has thyristor control instead of HT tap changing.
Traction Motors: GEC G412AZ frame mounted (87/0), G412BZ (87/1).
Max. Tractive Effort: 258 kN (58000 lbf).
Continuous Rating: 3730 kW (5000 hp) giving a tractive effort of 95 kN (21300 lbf) at 87 mph (Class 87/0), 3620 kW (4850 hp) giving a tractive effort of 96 kN (21600 lbf) at 84 mph (Class 87/1).
Maximum Rail Power: 5860 kW (7860 hp) at 50.8 mph.

Brake Force: 40 t.	**Length over Buffers:** 17.83 m.
Design Speed: 110 mph	**Weight:** 83.5 t.
Max. Speed: 110 (75*) mph.	**RA:** 6.
ETH Index: 95	**Wheel Diameter:** 1150 mm.
Train Brakes: Air.	**Electric Brake:** Rheostatic.

Class 87/0. Standard Design. Tap Changer Control.

87001	**I**	IWCA	WN	Royal Scot
87002	**I**	IWCA	WN	Royal Sovereign
87003	**I**	IWCA	WN	Patriot
87004	**I**	IWCA	WN	Britannia
87005	**I**	IWCA	WN	City of London
87006	**I0**	IWCA	WN	City of Glasgow
87007	**I**	IWCA	WN	City of Manchester
87008	**I**	IWCA	WN	City of Liverpool
87009	**I**	IWCA	WN	City of Birmingham
87010	**I**	IWCA	WN	King Arthur
87011	**I**	IWCA	WN	The Black Prince
87012	**M**	IWCA	WN	The Royal Bank of Scotland
87013	**I**	IWCA	WN	John O' Gaunt
87014	**I**	IWCA	WN	Knight of the Thistle
87015	**I**	IWCA	WN	Howard of Effingham
87016	**I**	IWCA	WN	Willesden Intercity Depot
87017	**I**	IWCA	WN	Iron Duke
87018	**I**	IWCA	WN	Lord Nelson
87019	**I**	IWCA	WN	Sir Winston Churchill
87020	**I**	IWCA	WN	North Briton
87021	**I**	IWCA	WN	Robert the Bruce
87022	**M**	IWCA	WN	Cock o' the North

87023	IO	IWCA	WN	Velocity
87024	I	IWCA	WN	Lord of the Isles
87025	IO	IWCA	WN	County of Cheshire
87026	I	IWCA	WN	Sir Richard Arkwright
87027	I	IWCA	WN	Wolf of Badenoch
87028	I	IWCA	WN	Lord President
87029	I	IWCA	WN	Earl Marischal
87030	I	IWCA	WN	Black Douglas
87031	M	IWCA	WN	Hal o' the Wynd
87032	IO	IWCA	WN	Kenilworth
87033	M	IWCA	WN	Thane of Fife
87034	IO	IWCA	WN	William Shakespeare
87035	M	IWCA	WN	Robert Burns

Class 87/1. Thyristor Control.

| 87101 | * | FD | DAMC | CE | STEPHENSON |

CLASS 90   GEC DESIGN   Bo – Bo

Built: 1987 – 90 by BREL at Crewe Works. Thyristor control.
Traction Motors: GEC G412CY separately excited frame mounted.
Max. Tractive Effort: 258 kN (58000 lbf).
Continuous Rating: 3730 kW (5000 hp) giving a tractive effort of 95 kN (21300 lbf) at 87 mph.
Maximum Rail Power: 5860 kW (7860 hp) at 50.8 mph.
Brake Force: 40 t.   **Length over Buffers:** 18.80 m.
Design Speed: 110 mph.   **Weight:** 84.5 t.
Max. Speed: 110 (75*) mph.   **RA:** 7.
ETH Index: 95   **Wheel Diameter:** 1156 mm.
Train Brakes: Air.   **Electric Brake:** Rheostatic.
Couplings: Drop-head buckeye.

Non-standard Liveries:

90128 is in SNCB/NMBS (Belgian Railways) electric loco livery.
90129 is in DB (German Federal Railways) 'neurot' livery.
90130 is in SNCF (French Railways) 'Sybic' livery.
90136 is in livery ''FE'', but with full yellow ends and roof and red 'Railfreight Distribution' lettering.

Class 90/0. As built.

90001	I	IWCA	WN	BBC Midlands Today
90002	I	IWCA	WN	The Girls' Brigade
90003	I	IWCA	WN	THE HERALD
90004	I	IWCA	WN	The D' Oyly Carte Opera Company
90005	I	IWCA	WN	Financial Times
90006	I	IWCA	WN	High Sheriff
90007	I	IWCA	WN	Lord Stamp
90008	I	IWCA	WN	The Birmingham Royal Ballet
90009	I	IWCA	WN	The Economist
90010	I	IWCA	WN	275 Railway Squadron (Volunteers)
90011	I	IWCA	WN	The Chartered Institute of Transport

90012	I	IWCA	WN	British Transport Police
90013	I	IWCA	WN	The Law Society
90014	I	IWCA	WN	'The Liverpool Phil'
90015	I	IWCA	WN	BBC North West
90016	RX	PXLA	CE	
90017	RX	PXLA	CE	
90018	RX	PXLA	CE	
90019	RX	PXLA	CE	Penny Black
90020	RX	PXLA	CE	Colonel Bill Cockburn CBE TD
90021	FD	DALC	CE	
90022	FD	DALC	CE	Freightconnection
90023	FD	DALC	CE	
90024	FD	DALC	CE	

Class 90/1. ETH equipment isolated. Renumbered from 90026 – 90150.

90125		FD	DAMC	CE	
90126	*	FD	DAMC	CE	Crewe Electric Depot Quality Approved
90127	*	FD	DAMC	CE	Allerton T&RS Depot Quality Approved
90128	*	O	DAMC	CE	Vrachtverbinding
90129	*	O	DAMC	CE	Frachtverbindungen
90130	*	O	DAMC	CE	Fretconnection
90131	*	FE	DAMC	CE	
90132	*	FE	DAMC	CE	Cerestar
90133	*	FE	DAMC	CE	
90134	*	M	DAMC	CE	
90135	*	M	DAMC	CE	
90136	*	O	DAMC	CE	
90137	*	FD	DAMC	CE	
90138	*	FD	DAMC	CE	
90139	*	FD	DAMC	CE	
90140	*	FD	DAMC	CE	
90141	*	FD	DAMC	CE	
90142	*	FD	DAMC	CE	
90143	*	FD	DAMC	CE	
90144	*	FD	DAMC	CE	
90145	*	FD	DAMC	CE	
90146	*	FD	DAMC	CE	
90147	*	FD	DAMC	CE	
90148	*	FD	DAMC	CE	
90149	*	FD	DAMC	CE	
90150	*	FD	DAMC	CE	

CLASS 91   GEC DESIGN   Bo – Bo

Built: 1988 onwards by BREL at Crewe Works. Thyristor control.
Traction Motors: GEC G426AZ.
Continuous Rating: 4540 kW (6090 hp).
Maximum Rail Power: 4700 kW (6300 hp).

Brake Force: 45 t.	**Length over Buffers:** 19.40 m.
Design Speed: 140 mph.	**Weight:** 84 t.
Max. Speed: 140 mph.	**RA:** 7.

ETH Index: 95   Wheel Diameter: 1000 mm.
Train Brakes: Air.   Electric Brake: Rheostatic.
Couplings: Drop-head buckeye.

91001	I	IECA	BN	Swallow
91002	I	IECA	BN	Durham Cathedral
91003	I	IECA	BN	THE SCOTSMAN
91004	I	IECA	BN	The Red Arrows
91005	I	IECA	BN	Royal Air Force Regiment
91006	I	IECA	BN	
91007	I	IECA	BN	Ian Allan
91008	I	IECA	BN	Thomas Cook
91009	I	IECA	BN	Saint Nicholas
91010	I	IECA	BN	
91011	I	IECA	BN	Terence Cuneo
91012	I	IECA	BN	
91013	I	IECA	BN	Michael Faraday
91014	I	IECA	BN	Northern Electric
91015	I	IECA	BN	
91016	I	IECA	BN	
91017	I	IECA	BN	Commonwealth Institute
91018	I	IECA	BN	Robert Louis Stevenson
91019	I	IECA	BN	Scottish Enterprise
91020	I	IECA	BN	
91021	I	IECA	BN	
91022	I	IECA	BN	Robert Adley
91023	I	IECA	BN	
91024	I	IECA	BN	
91025	I	IECA	BN	BBC Radio One FM
91026	I	IECA	BN	
91027	I	IECA	BN	
91028	I	IECA	BN	Guide Dog
91029	I	IECA	BN	Queen Elizabeth II
91030	I	IECA	BN	Palace of Holyroodhouse
91031	I	IECA	BN	Sir Henry Royce

CLASS 92   BRUSH DESIGN   Co–Co

Built: 1993 onwards by Brush Traction. Thyristor control.
Supply System: 25 kV a.c. from overhead equipment and 750 V d.c. third rail.
Electrical equipment: ABB Transportation, Zürich, Switzerland.
Traction Motors: Brush.
Max. Tractive Effort: 400 kN (90 000 lbf).
Continuous Rating at Motor Shaft: 5040 kW (6760 hp).
Maximum Rail Power: 5000 kW (6700 hp).
Brake Force: t.   **Length over Buffers:** 21.34 m.
Design Speed: 140 mph.   **Weight:** 126 t.
Max. Speed: 140 km/h (87.5 mph).   **RA:**
ETH Index: **Wheel Diameter:** 1160 mm.
Train Brakes: Air.   **Electric Brake:** Rheostatic.
Couplings: Drop-head buckeye.
Multiple Working: Time division multiplex system.

Communication Equipment: Driver – guard telephone and cab to shore radio-telephone.
Cab Signalling: Fitted with TVM430 cab signalling for Channel Tunnel.

f - Owned by SNCF.

92001	F	DAEC	CE	Victor Hugo
92002	F	DAEC	CE	H G Wells
92003	F	DAEC	CE	Beethoven
92004	F	DAEC	CE	Jane Austin
92005	F	DAEC	CE	Mozart
92006 f	F	DAEC	CE	Louis Armand
92007	F	DAEC	CE	Schubert
92008	F	DAEC	CE	Jules Verne
92009	F	DAEC	CE	Elgar
92010 f				Molière
92011				Handel
92012				Thomas Hardy
92013				Puccini
92014 f				Emile Zola
92015				D H Lawrence
92016				Brahms
92017				Shakespeare
92018 f				Stendhal
92019				Wagner
92020				Milton
92021				
92022				
92023				
92024				
92025				
92026				
92027				
92028				
92029				
92030				
92031				
92032				
92033				
92034				
92035				
92036				
92037				
92038				
92039				
92040				
92041				
92042				
92043				
92044				
92045				
92046				

SNCF CLASS 22200   B – B

Built: 1976 – 1986. by Alsthom/MTE. 22379/80 are converted from experimental locos 20011/2. These locos are included in this book because they work to England through the Channel Tunnel. The locos are labelled Tu (Trans-Manche – Unité multiple), or TTu (TVM430 + Tu) if fitted with cab signalling.
Supply System: 25 kV a.c./1500 V d.c. from overhead equipment.
Traction Motors: Two Alsthom monomotors.
Max. Tractive Effort: 294 kN (66 150 lbf).
Continuous Rating at Motor Shaft: 4360 kW (5900 hp).
Maximum Rail Power: 5000 kW (6700 hp).

Brake Force: t.	**Length over Buffers:** 17.48 m.
Design Speed: 200 mph.	**Weight:** 89 t.
Max. Speed: 160 km/h (100 mph).	**RA:**
ETH Index:	**Wheel Diameter:** 1250 mm.
Train Brakes: Air.	**Electric Brake:** Rheostatic.

Communication Equipment: Driver – guard telephone and cab to shore radio-telephone.
Multiple Working: Within class.
Non-standard Livery: SNCF Concrete & orange with yellow ends.

c Fitted with TVM430 cab signalling for Channel Tunnel.

22379	c	**0**	DASF	DP
22380	c	**0**	DASF	DP
22399	c	**0**	DASF	DP	MORMANT
22400		**0**	DASF	DP	MONTIGNY-EN-OSTREVENT
22401	c	**0**	DASF	DP	MOULINS
22402		**0**	DASF	DP	SAINT-DIE-DES-VOSGES
22403	c	**0**	DASF	DP	NEUVES-MAISONS
22404		**0**	DASF	DP	PAVILLONS-SOUS-BOIS
22405	c	**0**	DASF	DP	VILLIERS-SUR-MARNE

BR DEPARTMENTAL LOCOMOTIVES

CLASS 97/6   RUSTON SHUNTER   0-6-0

Built: 1959 by Ruston & Hornsby at Lincoln.
Engine: Ruston 6VPH of 123 kW (165 hp).
Main Generator: British Thomson Houston RTB6034.
Traction Motor: One British Thomson Houston RTA5041.
Max. Tractive Effort: 75 kN (17000 lbf).

Brake Force: 16 t.	**Length over Buffers:** 7.62 m.
Weight: 31 t.	**Wheel Diameter:** 978 mm.
Max. Speed: 20 mph.	**RA:** 1.

Non-Standard Livery: Departmental Yellow.

97651 (PWM 651)   v **0** LNCF   CF | 97654 (PWM 654)   v **0** EWOC   OC

CLASS 97/7   BATTERY LOCOS   Bo-Bo

Built: 1973 – 80 by BREL at Doncaster and Wolverton Works. Converted from Class 501 EMU cars.
Supply System: 750 V d.c. third rail or 320 V d.c. batteries.
Traction Motors: GEC WT344A.
Max. Tractive Effort: 73 kN (16400 lbf).

Brake Force: 45 t.	**Length over Buffers:** 18.44 m.
Weight: 59 t.	**Wheel Diameter:** 1071 mm.
Max. Speed: 25 mph.	**RA:** 4.

Multiple Working: Work in pairs.

97703	(61182)	a		NKFH	HE	97707	(61166)	a **N** NKFH HE
97704	(61185)	a		NKFH	HE	97708	(61173)	a **N** NKFH HE
97705	(61184)	a		NKFH	HE	97709	(61172)	a NKFH HE (S)
97706	(61189)	a		NKFH	HE	97710	(61175)	a NKFH HE (S)

CLASS 97/8   EE SHUNTER   0-6-0

For details see Class 09. Severn Tunnel emergency train locomotive.

Non-Standard Livery: BR blue with grey cab.

97806   (09017) xo   **0** LNCF   CF   Normally kept at Sudbrook.

DB 968xxx SERIES

This number series was introduced in 1969 and is for former capital stock locomotives which no longer operate under their own power. Departmental number not carried.

Non-standard Livery:

TDB 968030 (33018) x Fire Service Training Centre. Moreton-in-Marsh.

BR LOCOMOTIVES AWAITING DISPOSAL

03179	N	Ryde T&RSMD	08829		Toton TMD
08222		Bounds Green T&RSMD	08849		ABB Crewe Works
08239		Neville Hill T&RSMD	08855		Aberdeen T&RSMD
08305		Healey Mills	08870		Doncaster TMD
08309	F	Knottingley TMD	08880		Tinsley TMD
08393		Ipswich	08885		Doncaster TMD
08419		ABB Crewe Works	08895		Margam Wagon Works
08420		Port Talbot	08898		Bescot TMD
08439		Immingham TMD	20025		Frodingham
08468		Springs Branch	20028	BS	Falkland Junction
08478		Immingham TMD	20042		Frodingham
08507		Reading T&RSMD	20043		Frodingham
08515		Gateshead	20055		Falkland Junction
08537	FO	Bescot TMD	20071		Falkland Junction
08544		Heaton T&RSMD	20073		Springs Branch
08562		Stratford TMD	20082		Falkland Junction
08565		Motherwell TMD	20090	FR	Falkland Junction
08579		Healey Mills	20096		Thornaby TMD
08583		Doncaster TMD	20119		Toton
08590	BS	Heaton T&RSMD	20135		Toton
08595		Doncaster TMD	20142		Toton
08603		Stratford TMD	20151		Falkland Junction
08604	G	Derby T&RSMD	20154		Toton
08609		Willesden TMD	20172		Falkland Junction
08618		Gateshead	20177		Toton
08627		Ipswich	20186		Falkland Junction
08631	N	March TMD	20195		Falkland Junction
08634		Stratford TMD	25083		Basford Hall Yard
08647	G	ABB Crewe Works	25205		Bescot Yard
08659		Healey Mills	25211		Bescot Yard
08667		Neville Hill T&RSMD	25259		Bescot Yard
08677		Willesden TMD	26001	G	Inverness T&RSMD
08684		Bletchley TMD	26002	FC	Inverness T&RSMD
08692		ABB Crewe Works	26003	C	Inverness T&RSMD
08699		ABB Crewe Works	26004	C	Inverness T&RSMD
08733		Motherwell TMD	26005	C	Inverness T&RSMD
08748		Ipswich	26006	FC	Inverness T&RSMD
08755		Millerhill	26007	G	Inverness T&RSMD
08760		BRML Eastleigh Works	26010	FR	Inverness T&RSMD
08767		Ipswich	26011	C	Motherwell TMD
08771		Heaton T&RSMD	26014		Perth
08772	G	Colchester	26024		Motherwell TMD
08777		Hull Botanic Gardens FP	26025	C	Inverness T&RSMD
08788		Derby T&RSMD	26026	C	Perth
08789		Bletchley TMD	26027		Perth
08793	O	Aberdeen T&RSMD	26032	FR	Inverness T&RSMD
08794		Neville Hill T&RSMD	26035	C	Inverness T&RSMD

26036	C	Inverness T&RSMD	37008	FR	ABB Crewe Works
26037	FR	Inverness T&RSMD	37029	FD	Crewe Electric TMD
26038	FR	Inverness T&RSMD	37031	FD	Cardiff Canton T&RSMD
26040	C	Perth	37032	FD	Doncaster TMD
26041	FR	Inverness T&RSMD	37190	FM	Gateshead
26042		Inverness T&RSMD	37215	FP	Inverness T&RSMD
26043	C	Perth	37373	FR	Old Oak Common TMD
31108	FO	Scunthorpe Yard	37681	FA	ABB Crewe Works
31123		Bescot Yard	45015		Toton
31156		Scunthorpe Yard	47008		Stratford DRS
31168		Bescot Yard	47019	FO	Crewe Electric TMD
31196	C	Stratford TMD	47094	FP	Scunthorpe Yard
31210	FO	Scunthorpe Yard	47108		Old Oak Common TMD
31212		Scunthorpe Yard	47112	FO	Old Oak Common TMD
31215	FO	Immingham TMD	47115		Scunthorpe Yard
31217	FC	Toton	47118	BR	Healey Mills
31221		Scunthorpe Yard	47119	FP	Immingham TMD
31223		Immingham TMD	47143		Doncaster TMD
31240	FO	Stratford TMD	47198		Cardiff Canton T&RSMD
31243	FO	Stratford TMD	47233	FP	Scunthorpe Yard
31249		Immingham TMD	47318	FO	Bescot TMD
31264		Thornaby TMD	47320	FO	ABB Crewe Works
31282	FR	Crewe Diesel TMD	47373	FP	Scunthorpe Yard
31283	0	Stratford TMD	47380	FP	Scunthorpe Yard
31286		Bescot TMD	47381	FP	Scunthorpe Yard
31289		Bescot Yard	47406	I0	Scunthorpe Yard
31293		Stratford TMD	47407	BR	Scunthorpe Yard
31296	FA	Crewe Diesel TMD	47411	BR	Scunthorpe Yard
31299	FO	Stratford TMD	47413	BR	Scunthorpe Yard
31305		Bescot TMD	47417		Scunthorpe Yard
31320		Stratford TMD	47418		Scunthorpe Yard
31402		Toton	47421		Crewe Diesel TMD
31428		Basford Hall Yard	47423		Old Oak Common TMD
31442		Crewe Diesel TMD	47425		Old Oak Common TMD
31460		Bescot TMD	47426	BR	Old Oak Common TMD
31970	0	ABB Crewe Works	47430	FA	Old Oak Common TMD
33009	C	BRML Eastleigh Works	47431	BR	Old Oak Common TMD
33020		Stewarts Lane T&RSMD	47432	BR	Holbeck
33033	FA	Stewarts Lane T&RSMD	47433	BR	Crewe Diesel TMD
33038		Stratford TMD	47438	BR	Old Oak Common TMD
33040		Stewarts Lane T&RSMD	47439	BR	Crewe Diesel TMD
33047	C	Eastleigh Yard	47440	BR	Old Oak Common TMD
33050	FA	Stewarts Lane T&RSMD	47441	BR	Old Oak Common TMD
33101	D	Eastleigh T&RSMD	47442	BR	Crewe Diesel TMD
33108	C	Eastleigh T&RSMD	47443	BR	Crewe Diesel TMD
33113		Stewarts Lane T&RSMD	47444	BR	Basford Hall Yard
33114	N	Eastleigh T&RSMD	47446	BR	Old Oak Common TMD
33118	C	Eastleigh T&RSMD	47448	BR	Holbeck
33201	C	Stewarts Lane T&RSMD	47452	BR	Old Oak Common TMD
33205	FD	Stewarts Lane T&RSMD	47453	BR	Old Oak Common TMD
33211	FD	Stewarts Lane T&RSMD	47455	BR	ABB Crewe Works

47457	**BR**	Old Oak Common TMD	
47458	**R**	Holbeck	
47465	**BR**	Old Oak Common TMD	
47466	**BR**	ABB Crewe Works	
47470	**M**	ABB Crewe Works	
47472		Old Oak Common TMD	
47482	**BR**	ABB Crewe Works	
47483	**M**	Crewe Diesel TMD	
47485	**BR**	Crewe Diesel TMD	
47508	**M**	Bristol Bath Road TMD	
47509	**I**	Bristol Bath Road TMD	
47515	**M**	Holbeck	
47518	**BR**	Immingham TMD	
47527	**M**	Bristol Bath Road TMD	
47533	**R**	Old Oak Common TMD	
47534	**BR**	Longsight T&RSMD	

47538	**BR**	York
47542		Stratford TMD
47549	**IO**	Crewe Diesel TMD
47633	**BR**	BRML Glasgow Works
47643	**IO**	Inverness T&RSMD
50029	**N**	Laira T&RSMD
50030	**N**	Laira T&RSMD
56013	**FC**	Toton
56015	**FC**	BRML Doncaster Works
56023	**FC**	Toton
56028	**FC**	Toton TMD
56030	**FC**	Toton TMD
56122	**FC**	Toton
73004	**0**	Birkenhead N. T&RSMD
73111	**IO**	Stewarts Lane T&RSMD
97653	**0**	Reading T&RSMD

Non-Standard Liveries:

08793 London & North Eastern Railway apple green
31283 BR blue with large numbers
31970 Research light grey, dark grey, white and red
73004 NSE blue
97653 Departmental yellow

Network SouthEast liveried Class 47 No. 47711 'County of Hertfordshire' hauls a pair of condemned 4 EPB units through the outskirts of Bristol en-route from Bournemouth to Margam for scrapping on 10th March 1994.

Nic Joynson

▲ Recently repainted in BR blue, Class 56 No. 56004 is pictured stabled at Wigan Springs Branch on 4th April 1994.   *Paul Senior*

▼ Trainload Construction liveried Class 56 No. 56103 hauls a Wisbech – Glasgow Deanside pet food train through Cowran cutting, near Brampton on 16th May 1994.   *Kevin Conkey*

Trainload Coal liveried Class 58 No. 58045 passes Polhill, just north of Sevenoaks, with a training trip from Hoo Junction on 7th February 1994.   *Chris Wilson*

Class 60 No. 60029 'Ben Nevis', in Trainload Metals livery, passes through the centre of Poole with a Hamworthy – Cardiff Tidal sidings steel train on 28th January 1994.
Nic Joynson

▲ Class 73 No. 73006 is seen carrying the new Merseyrail departmental livery whilst stabled at Hall Road EMU Depot on 10th April 1994.
M Hilbert

▼ Gatwick Express liveried Class 73 No. 73212 'Airtour Suisse' is seen passing Clapham Junction whilst working the 14.45 London Victoria – Gatwick Airport. *Hugh Ballantyne*

▲ Rail Express Systems liveried Class 86 No. 86261 passes through Lower Hatton, North Staffordshire with the 15.30 London Euston to Crewe Postal on 8th June 1993.   *Hugh Ballantyne*

▼ Old InterCity liveried Class 87 No. 87006 'City of Glasgow' is pictured providing the power for the 16.10 London Euston – Liverpool Lime Street service on 30th April 1994.   *Brian Denton*

A Silcock Express Dagenham – Garston car train is seen near Leighton Buzzard on 29th March 1994 being hauled by new Railfreight Distribution liveried Class 90 No. 90133.   *Paul D Shannon*

▲ Class 91 No. 91015, in InterCity livery, passes Blindwells on 13th July 1992 with a down East Coast Main Line service. *Paul D Shannon*

▼ SNCF Class 22200 Nos. 22400 & 22399 at Lille la Delivrance yard during trials in October 1993. *David Haydock*

POOL CODES & ALLOCATIONS

CENTRAL SERVICES

CDJC   Crewe Diesel Class 47 (Research)

47971 BR 47972 CS 47973 M 47975 C 47976 C 47981 C

CDJD   Derby Etches Park Class 08 (Research)

08417 D 08956

RAILFREIGHT DISTRIBUTION

DAAN   Allerton Class 08

08482 D	08569	08585	08624	08694	08703	08737 F
08739	08784	08799	08856	08872 D	08891	08902
08907 O	08939	08951 D				

DACT   Tinsley Class 47 (Channel Tunnel)

47053 FE 47085 FE 47125 FE 47186 FE 47217 FE 47234 FE 47245 FE
47286 FE 47290 FE 47299 FE 47307 FE 47344 FE 47351 FE 47365 FE

DADR   Tinsley Class 09 (Dover)

08653   08825   08837 D 08913 D

DAEC   Crewe Electric Class 92

92001 F 92002 F 92003 F 92004 F 92005 F 92006 F 92007 F
92008 F 92009 F

DALC   Crewe Electric Class 90/0

90021 FD 90022 FD 90023 FD 90024 FD

DAMC   Crewe Electric Class 87/1 & 90/1

87101 FD 90125 FD 90126 FD 90127 FD 90128 O 90129 O 90130 O
90131 FD 90132 FE 90133 FE 90134 M 90135 M 90136 O 90137 FD
90138 FD 90139 FD 90140 FD 90141 FD 90142 FD 90143 FD 90144 FD
90145 FD 90146 FD 90147 FD 90148 FD 90149 FD 90150 FD

DAMT   Tinsley Class 37/47 (Felixstowe)

37131 FD 37178 FD 37225 FD 37298 FD 47390 FD 47391 FD 47392 FD
47393 FD 47394 FD 47395 FD 47396 FD 47397 F 47398 FD 47399 FD

DANC   Crewe Electric Class 86/6

86602 FD 86603 FD 86604 FD 86605 FD 86606 FD 86607 FD 86608 FE
86609 FD 86610 FD 86611 FD 86612 FD 86613 FD 86614 FD 86615 FD
86618 FD 86620 FD 86621 FD 86622 FD 86623 FD 86627 FD 86628 FD
86631 FE 86632 FD 86633 FD 86634 FD 86635 FD 86636 FD 86637 FD
86638 FD 86639 FD

DART   Tinsley Class 47 (Restricted)

47142 **FR** 47157 **F** 47249 **FR** 47270   47301 **FR** 47317 **FD** 47322 **FR**
47345 **FR** 47350 **FO** 47358 **FO** 47371 **FO**

DASF   Dijon Perrigny SNCF Class 22200 (Channel Tunnel)

22379 **O** 22380 **O** 22399 **O** 22400 **O** 22401 **O** 22402 **O** 22403 **O**
22404 **O** 22405 **O**

DAST   Tinsley Class 47 (Single Tank)

47052 **FD** 47060 **FD** 47145 **O** 47146   47147 **FD** 47187 **FD** 47206 **FD**
47225 **FD** 47231 **FD** 47279 **FD** 47283 **FD** 47296 **FD** 47305 **FP** 47337 **FO**
47339 **FD** 47347 **FM** 47349 **FD** 47354 **FD** 47367 **FR** 47377 **FD**

DASY   Tinsley Class 08 (Saltley)

08413 **D** 08535 **D** 08751 **FE** 08905   08946 **D**

DATI   Tinsley Class 08

08389   08575 **BS** 08745 **BS** 08879

DATT   Tinsley Class 47 (Twin Tank)

47049 **FD** 47095 **FD** 47114 **FD** 47144 **FD** 47156 **FD** 47188 **FD** 47194 **FD**
47201 **FD** 47210 **FD** 47213 **FD** 47218 **FD** 47226 **FD** 47228 **FD** 47229 **FD**
47258 **FD** 47281 **FD** 47284 **FD** 47287 **FD** 47291 **FD** 47292 **FD** 47293 **FD**
47297 **FD** 47298 **FD** 47306 **FE** 47312 **FD** 47335 **FD** 47360 **FD** 47361 **FD**
47363 **F** 47375 **FE** 47378 **FD** 47387 **FD** 47388 **FD** 47389 **FD**

DAUT   Tinsley Class 47 (Automotive)

47033 **FD** 47051 **FD** 47200 **FD** 47219 **FD** 47222 **FD** 47236 **FD** 47237 **FD**
47241 **FE** 47280 **FD** 47285 **FD** 47310 **FD** 47313 **FD** 47316 **FD** 47323 **FE**
47326 **FD** 47338 **FE** 47362 **FD**

DAWE   Tinsley Class 08 (Wembley)

08530 **D** 08531 **D** 08642 **O** 08655 **F** 08842   08892 **D** 08926
09011 **D** 09021   09022

DAYX   Stored Locos

08661   08673 **IO** 08691 **G** 37110 **FD** 37218 **FD** 37238 **FD** 47050 **FD**
47063 **FA** 47190 **FP** 47214 **FD** 47288 **FD** 47289 **FD** 47321 **F** 47325 **FO**
47376

TRAINLOAD FREIGHT SOUTH EAST

ENAN   Toton Class 60

60006 **FA** 60009 **FA** 60010 **FA** 60011 **FA** 60012 **FA** 60017 **FA** 60044 **FM**
60048 **FA** 60071 **F** 60072 **FC** 60073 **FC** 60074 **FC** 60075 **FC** 60076 **FC**
60077 **FC** 60078 **FC** 60079 **F** 60083 **FA** 60086 **FC** 60087 **FC** 60088 **FC**
60094 **FA** 60098 **FA**

ENBN   Toton Class 58

58001 **FC** 58002 **F** 58003 **F** 58004 **FC** 58005 **F** 58006 **FC** 58007 **FC**
58008 **FC** 58009 **FC** 58010 **FC** 58011 **F** 58012 **F** 58013 **FC** 58014 **F**
58015 **F** 58016 **FC** 58017 **FC** 58018 **FC** 58019 **F** 58020 **FC** 58021 **FC**
58022 **F** 58023 **FC** 58024 **FC** 58025 **F** 58026 **F** 58027 **FC** 58028 **FC**
58029 **FC** 58030 **F** 58031 **F** 58032 **F** 58033 **FC** 58034 **FC** 58035 **F**
58036 **FC** 58037 **FC** 58038 **FC** 58039 **FC** 58040 **F** 58041 **F** 58042 **FC**
58043 **FC** 58044 **FC** 58045 **FC** 58046 **FC** 58047 **FC** 58048 **FC** 58049 **FC**
58050 **FC**

ENDN   Toton Class 31/37 (Derby Infrastructure)

31403   31407 **M** 31459   31461 **D** 37057 **BR** 37092 **C** 37114 **C**
37185 **C**

ENPN   Toton Class 31 (Peterborough Infrastructure)

31116 **O** 31135 **C** 31250 **C** 31271 **FA** 31276 **FC** 31308 **C** 31531 **C**
31541 **C** 31549 **C** 31551 **C** 31552 **C** 31558 **C**

ENRN   Toton Class 31 (Restricted)

31149 **FR** 31165 **G** 31180 **FR** 31181 **C** 31186 **C** 31187 **C** 31191 **C**
31205 **FR** 31219 **C** 31230 **FO** 31247 **FR** 31268 **C** 31290 **C** 31294 **FA**
31466 **C** 31547 **C** 31553 **C** 31563 **C** 31569 **C**

ENSN   Toton Class 08 (Toton/Peterborough)

08441   08449   08492   08495   08511   08528 **D** 08529
08538 **D** 08540 **D** 08580   08597   08607   08723   09201 **D**

ENSX   Toton Class 08 (Stored)

08773

ENXX   Stored Locos

31184 **FO** 31209 **FA** 31252 **FO** 37278 **FC**

ENZX   Locos For Withdrawal

33064 **FA** 73003 **G**

ESAB   Stewarts Lane Class 60

60001 **FA** 60018 **FA** 60019 **FA** 60039 **FA** 60040 **FA** 60041 **FA** 60042 **FA**
60043 **FA** 60099 **FA** 60100 **FA**

ESBB   Stewarts Lane Class 37/7

37703 **FC** 37705 **FP** 37709 **FP** 37715 **FP** 37798 **FC** 37800 **FC** 37803 **FC**
37890 **FP** 37891 **FP** 37892 **FP**

ESPS   Stratford Class 37/5 (Anglia Petroleum)

37667 **FP** 37676 **FA** 37678 **FA** 37679 **FA**

EWAS   Stratford Class 47

47223 **FD** 47278 **FP** 47315 **C** 47368 **FP** 47462 **R** 47674 **BR**

EWCN   Toton Class 37 (GW Infrastructure)

37010 **C** 37012 **C** 37035 **C** 37038 **C** 37040 **FM** 37042 **FM** 37046 **C**
37048 **FM** 37072 **D** 37097 **C** 37098 **C** 37137 **FM** 37162 **D** 37174 **C**
37203 **FM** 37213 **FC** 37222 **FC** 37227 **FM** 37264 **C**

EWDB   Stewarts Lane Class 33/37 (Infrastructure)

33002 **C** 33008 **G** 33019 **C** 33025 **C** 33026 **C** 33030 **C** 33035 **N**
33046 **C** 33051 **C** 33057 **C** 33065 **C** 33109 **D** 33116   33202 **C**
33204 **FD** 33206 **FD** 33207 **FA** 33208 **C** 37198 **C** 37274 **C** 37375 **C**
37377 **C**

EWDS   Stratford Class 37 (Anglia Infrastructure)

37013 **F** 37023 **C** 37047 **FD** 37054 **C** 37055 **FD** 37077 **FM** 37106 **C**
37109 **FM** 37140 **C** 37167 **FC** 37216 **G** 37219   37241 **FM** 37242 **FD**
37244 **FC** 37280 **FP** 37370 **C** 37371 **C** 37376 **C** 37379 **C**

EWEB   Stewarts Lane Class 73 (Infrastructure)

73103 **IO** 73104 **IO** 73105 **C** 73106 **D** 73108 **C** 73110 **C** 73114 **IO**
73117 **IO** 73119 **C** 73130 **C** 73133 **N** 73134 **IO** 73136 **N** 73138 **C**

EWEH   Eastleigh Class 08

08600 **D** 08854   08933 **O** 08940

EWHB   Stewarts Lane Class 73 (For Hire)

73101 **O** 73107 **C** 73118 **C** 73129 **N**

EWOC   Old Oak Common Class 08/09/97

08460 **F** 08480 **G** 08523   08646 **F** 08651 **D** 08664   08904
08924 **D** 08944 **D** 08947   09101 **D** 09102 **D** 97654 **O**

EWRB   Stewarts Lane Class 33/37/73 (Restricted)

33012   33021 **FA** 33023   33029   33042 **FA** 33048   33052
33053 **FA** 33063 **C** 33103 **C** 33117   37194 **FD** 37220 **FP** 37245 **C**
37293 **FM** 37380 **FC** 73126 **N** 73128 **C** 73131 **C** 73132 **IO** 73139 **IO**
73140 **IO** 73141 **IO**

EWRN   Toton Class 37 (Restricted)

37065 **FD** 37070 **FD** 37074 **FD** 37138 **FM** 37372 **C**

EWRS   Stratford Class 47 (Restricted)

47004 **G** 47016 **FO** 47121   47348 **FO** 47366 **FO** 47484 **G** 47526 **BR**
47802 **I** 47803 **O** 47804 **I**

EWSF   Stratford Class 08

08414 **O** 08526   08541 **D** 08542 **F** 08593 **O** 08670   08689 **O**
08709   08715 **O** 08740 **F** 08750   08752 **C** 08758   08775
08828   08866   08909   08957

EWSU   Selhurst Class 08/09

08698   09003   09006   09007   09009 **D** 09010 **D** 09012 **D**
09016 **D** 09018   09019 **D** 09020   09023   09024 **D**

EWSX   **Stratford Class 08 (Stored)**

08517   08700   08811   08878   08958

EWTS   **Stratford Class 47 (New Stock Delivery)**

47579 **N** 47702 **N** 47711 **N**

TRAINLOAD FREIGHT NORTH EAST

FDAI   Immingham Class 60 (Humberside)

60003 **FP** 60004 **FC** 60008 **FM** 60013 **FP** 60014 **FP** 60021 **FM** 60024 **FP**
60025 **FP** 60026 **FP** 60028 **FP** 60050 **F**   60051 **FP** 60053 **FP** 60054 **FP**
60091 **FC**

FDAK   Immingham Class 60 (Aire Valley)

60002 **FP** 60027 **FP** 60059 **FC** 60064 **FP** 60067 **F**   60068 **F**   60069 **F**
60070 **FC**

FDBI   Immingham Class 56 (Humberside)

56035 **FA** 56048 **C** 56051 **FA** 56061 **FM** 56084 **FC** 56085 **FC** 56088 **FC**
56089 **FC** 56090 **FC** 56094 **FC** 56106 **FC** 56126 **FC**

FDBK   Immingham Class 56 (Aire Valley)

56005 **FC** 56006 **FC** 56011 **F** 56021 **FC** 56031 **C** 56041 **FA** 56043 **F**
56046 **C** 56055 **FA** 56067 **FC** 56068 **FC** 56074 **FC** 56075 **F** 56077 **FC**
56078 **F** 56080 **F** 56082 **FC** 56083 **FC** 56091 **F** 56095 **F** 56098 **FC**
56100 **FC** 56102 **F**

FDCI   Immingham Class 37 (Humberside)

37512 **FM** 37515 **FM** 37517 **FM** 37519 **FM** 37677 **F** 37680 **FA** 37684 **FA**
37688 **FA** 37689 **F** 37694 **FC** 37698 **FC** 37699 **FC** 37706 **FP** 37707 **FP**
37708 **FP** 37710 **FP** 37711 **FM** 37713 **FM** 37717 **FP** 37719 **FP** 37883 **FP**
37884 **FP** 37885 **FP** 37886 **FM**

FDDI   Immingham Class 37/47 (Doncaster Departmental)

37003 **C** 37009 **FD** 37049 **C** 37058 **C** 37083 **C** 37095 **C** 47197 **FP**
47212 **FP** 47221 **FP** 47224 **FP** 47277 **FD** 47294 **FD** 47676 **I**   47677 **I**

FDEI   Immingham Class 47 (Electrification)

47319 **FP** 47331 **C** 47346 **C** 47369 **FD**

FDRI   Immingham Class 37/47 (Restricted)

37079 **FD** 37139 **FC** 37223 **FC** 37235 **F** 37252 **FD** 37271 **FD** 47256 **FD**
47276 **FP** 47352 **C** 47359 **FD** 47550 **M**

FDSD   Doncaster Class 08

08418 **F** 08500 **O** 08509 **F** 08510   08512 **F** 08514   08581 **BS**
08713   08813 **D** 08824 **F** 08877 **D** 08903

FDSI   Immingham Class 08

08388 **FP** 08401 **D** 08405 **D** 08466 **FO** 08632   08665

FDSK   Knottingley Class 08/09

08442 F 08499 F 08516 D 08605   08662   08706   08776 D
08782   08783   09005 D 09014 D

FDSX   Immingham Class 08

08445

FDYX   Stored Locos

37101 FD 37104 C 37144 FA 37209 BR 37381 FD 37382 FP 37888 FP
47379 FP 47555 I0 56003 F 56008   56012 FC 56014 FC 56024 F0
56026   56027 FC

FEPS   Immingham/Thornaby Class 37/5 (For EPS)

37501 FM 37502 FM 37504 FM 37506 FM 37507 FM 37508 FM 37511 FM
37513 FM 37514 FM 37687 FA 37690 F0 37691 F0

FMAY   Thornaby Class 60

60007 FP 60020 FM 60022 FM 60023 FM 60030 FM 60031 FM 60038 FM
60049 FM 60052 FM 60090 FC

FMBB   Immingham Class 56 (Blyth)

56108 F 56109 FC 56110 FA 56111 FC 56112 FC 56117 FC 56118 FC
56120 FC 56130 FC 56131 F 56134 FC 56135 F

FMBY   Thornaby Class 56

56034 FA 56039 FA 56045 FA 56050 FA 56062 F 56063 FA 56065 FA
56069 FM 56081 F 56087 FM 56097 FM 56107 FC 56116 FC

FMCY   Thornaby Class 37 (Refurbished)

37415 M 37419 M 37426 M 37516 FM 37682 FA 37697 FC 37716 FM
37718 FM

FMDY   Thornaby Class 37 (Departmental)

37053 FD 37068 FD 37075 F 37350 FP 37358 F 37359 FP 37378 FD

FMRY   Thornaby Class 37 (Restricted)

37015 FD 37019 FD 37045 F 37059 FD 37063 FD 37128 BR 37202 FM
37217   37239 FC 37272 FD 37285 F

FMSY   Thornaby Class 08/09

08577   08582 D 08587   08806 F 08931   09106 D 09204 D

EUROPEAN PASSENGER SERVICES

GPSS   Old Oak Common Class 08 (North Pole)

08948

TRAIN OPERATING UNITS

HASS   ScotRail TOU - Inverness Class 08
08754   08762

HBSH   ECML TOU - Bounds Green/Edinburgh Craigentinny Class 08
08472   08571   08724   08834 **FD** 08853

HEBD   Merseyrail TOU - Birkenhead North Class 73/0
73001 **MD** 73002 **BR** 73005 **O** 73006 **MD**

HFSL   WCML TOU - Longsight Class 08
08611   08721 **O** 08790

HFSN   WCML TOU - Willesden Class 08
08451   08454   08617   08696 **D** 08887   08934

HGSS   Central TOU - Tyseley Class 08
08616   08805 **FO**

HISE   Midland/Cross Country TOU - Derby Etches Park Class 08
08536   08690   08697   08899

HISL   Midland/Cross Country TOU - Neville Hill Class 08
08525 **F** 08588 **BS** 08908   08950 **I**

HJSA   Great Western TOU - Bristol Bath Road Class 08
08410 **D** 08483 **D** 08643 **D** 08836

HJSE   Great Western TOU - Landore Class 08
08780   08795 **D** 08822

HJSL   Great Western TOU - Laira Class 08
08641 **D** 08644 **I** 08645 **D** 08648 **D** 08663 **D**

HLSV   Cardiff Valleys TOU - Cardiff Canton Class 08
08830

HSSN   Anglia TOU - Norwich Crown Point Class 08
08810   08869 **G** 08928 **FR**

HWSU   South Central TOU - Selhurst Class 09
09004   09025   09026 **D**

HYSB   South Western TOU - Bournemouth Class 73/1
73109 **N**

HZSH   Isle Of Wight TOU - Ryde Class 03
03079

INTERCITY TRAIN OPERATING UNITS

Please note that InterCity locos are now owned by three train leasing companies (LEASCOS). These are:

SA - Eversholt Train Leasing
SB - Porter Brook Train Leasing
SC - Angel Train Leasing

The codes are shown in brackets after the pool details

IANA   Anglia TOU - Norwich Crown Point Class 86/2 (SA)

| 86215 I | 86217 I | 86218 I | 86220 I | 86221 I | 86223 I | 86230 I |
| 86232 I | 86235 I | 86237 I | 86238 I | 86246 I | 86250 I | |

ICCA   Cross Country TOU - Longsight Class 86/2 (SA)

86205 I	86206 I	86212 I	86214 I	86216 I	86222 I	86226 M
86227 M	86228 I	86229 I	86233 I	86234 I	86244 I	86247 I
86252 I	86255 I	86259 I	86260 I			

ICCP   Cross Country TOU - Neville Hill/Laira Class 43 (SB)

43086 I	43087 I	43088 I	43089 I	43101 I	43102 I	43103 I
43121 I	43122 I	43153 I	43154 I	43155 I	43156 I	43157 I
43158 I	43159 I	43160 I	43161 I	43162 I	43178 I	43180 I
43184 I	43193 I	43194 I	43195 I	43196 I	43197 I	43198 I

ICCS   Cross Country TOU - Edinburgh Craigentinny Class 43 (SB)

43006 I	43007 I	43008 I	43013 I	43014 I	43062 I	43063 I
43065 I	43067 I	43068 I	43069 I	43070 I	43071 I	43078 I
43079 I	43080 I	43084 I	43090 I	43091 I	43092 I	43093 I
43094 I	43097 I	43098 I	43099 I	43100 I	43123 I	

IECA   ECML TOU - Bounds Green Class 91 (SA)

91001 I	91002 I	91003 I	91004 I	91005 I	91006 I	91007 I
91008 I	91009 I	91010 I	91011 I	91012 I	91013 I	91014 I
91015 I	91016 I	91017 I	91018 I	91019 I	91020 I	91021 I
91022 I	91023 I	91024 I	91025 I	91026 I	91027 I	91028 I
91029 I	91030 I	91031 I				

IECP   ECML TOU - Edinburgh Craigentinny/Neville Hill Class 43 (SC)

43038 I	43039 I	43095 I	43096 I	43104 I	43105 I	43106 I
43107 I	43108 I	43109 I	43110 I	43111 I	43112 I	43113 I
43114 I	43115 I	43116 I	43117 I	43118 I	43119 I	43120 I

ILRA   Cross Country TOU - Bristol Bath Road Class 47/8 (SB)

47805 I	47806 I	47807 I	47810 I	47811 I	47812 I	47813 I
47814 I	47815 I	47816 I	47817 I	47818 I	47822 I	47825 I
47826 I	47827 I	47828 I	47829 I	47830 I	47831 I	47832 I
47839 I	47840 I	47841 I	47843 I	47844 I	47845 I	47846 I
47847 I	47848 I	47849 M	47850 I	47851 I	47853 M	

IMLP   Midland Main Line TOU - Neville Hill Class 43 (SB)

43043 I	43044 I	43045 I	43046 I	43047 I	43048 I	43049 I
43050 I	43051 I	43052 I	43053 I	43054 I	43055 I	43056 I
43057 I	43058 I	43059 I	43060 I	43061 I	43064 I	43066 I
43072 I	43073 I	43074 I	43075 I	43076 I	43077 I	43081 I
43082 I	43083 I	43085 I				

IVGA   Gatwick Express TOU - Stewarts Lane Class 73 (SB)

73112 N	73201 I	73202 GE	73203 I	73204 GE	73205 M	73206 GE
73207 GE	73208 GE	73209 GE	73210 GE	73211 I	73212 GE	73235 GE

IWCA   WCML TOU - Willesden Class 87/90 (SB)

87001 I	87002 I	87003 I	87004 I	87005 I	87006 IO	87007 I
87008 I	87009 I	87010 I	87011 I	87012 M	87013 I	87014 I
87015 I	87016 I	87017 I	87018 I	87019 I	87020 I	87021 I
87022 M	87023 IO	87024 I	87025 IO	87026 I	87027 I	87028 I
87029 I	87030 I	87031 M	87032 IO	87033 M	87034 IO	87035 M
90001 I	90002 I	90003 I	90004 I	90005 I	90006 I	90007 I
90008 I	90009 I	90010 I	90011 I	90012 I	90013 I	90014 I
90015 I						

IWPA   WCML TOU - Willesden Class 86/2 (SA)

86101 I	86102 IO	86103 I	86204 I	86207 I	86209 M	86213 I
86219 I	86224 I	86225 I	86231 I	86236 I	86240 I	86242 I
86245 I	86248 I	86249 M	86251 I	86253 I	86256 I	86257 I
86258 I						

IWRP   GWML TOU - Laira/St Phillips Marsh Class 43 (SC)

43002 I	43003 I	43004 I	43005 I	43009 I	43010 I	43011 I
43012 I	43015 I	43016 I	43017 I	43018 I	43019 I	43020 I
43021 I	43022 I	43023 I	43024 I	43025 I	43026 I	43027 I
43028 I	43029 I	43030 I	43031 I	43032 I	43033 I	43034 I
43035 I	43036 I	43037 I	43040 I	43041 I	43042 I	43124 I
43125 I	43126 I	43127 I	43128 I	43129 I	43130 I	43131 I
43132 I	43133 I	43134 I	43135 I	43136 I	43137 I	43138 I
43139 I	43140 I	43141 I	43142 I	43143 I	43144 I	43145 I
43146 I	43147 I	43148 I	43149 I	43150 I	43151 I	43152 I
43163 I	43164 I	43165 I	43166 I	43168 I	43169 I	43170 I
43171 I	43172 I	43173 I	43174 I	43175 I	43176 I	43177 I
43179 I	43181 I	43182 I	43183 I	43185 I	43186 I	43187 I
43188 I	43189 I	43190 I	43191 I	43192 I		

IXXA   Stored Locos

43167 I

BRML & MAJOR DEPOT SHUNTERS

KCSI Ilford Level 5 Depot Class 08

08527 **D** 08573

KDSD BRML Doncaster Class 08

08682 08823

KESE BRML Eastleigh Class 08

08649 **G** 08847

KGSS BRML Springburn Class 08

08568 08730 **O**

KWSW BRML Wolverton Class 08

08484 **D** 08629 **O**

TRAINLOAD FREIGHT WEST

LBBS Bescot Class 08

08428　　08543 **D** 08601 **O** 08623　　08734　　08746 **D** 08765 **D**
08914　　08920 **F** 09104 **D**

LBBX Bescot Class 08 (Stored)

08448　　08610　　08893 **D** 08901

LBBY Bletchley Class 08

08519 **O** 08567　　08625　　08628　　08683　　08807 **BS** 08927

LBCB Bescot Class 47 (Departmental)

47193 **FP** 47295 **FP** 47302 **FR** 47308 **F** 47329 **C** 47332 **C** 47333 **C**
47334 **C** 47340 **C** 47341 **C** 47353 **C** 47372 **C** 47473 **BR** 47478

LBDB Bescot Class 31 (Departmental)

31102 **C** 31105 **C** 31106 **C** 31107 **C** 31110 **C** 31112 **C** 31113 **C**
31125 **C** 31146 **C** 31147 **C** 31155 **FA** 31166 **C** 31178 **C** 31185 **C**
31273 **C** 31405 **M** 31417 **D** 31420 **M** 31422 **M** 31423 **M**
31434　　31435 **C** 31450　　31462 **D** 31467　　31468 **C** 31524 **C**
31530 **C** 31537 **C** 31545　　31546 **C** 31554 **C**

LBHB Bescot Class 20

20016　　20057　　20059 **FR** 20066　　20081　　20087 **BS** 20092 **CS**
20118 **FR** 20132 **FR** 20138 **FR** 20165 **FR** 20168　　20169 **CS**

LBRB Bescot Class 31/47 (Restricted)

31128 **FO** 31132 **FO** 31164 **FO** 31206 **C** 31232 **C** 31317 **FO** 31411 **D**
31415　　31514 **C** 31526 **C** 31533 **C** 31548 **C** 47079 **FD** 47207 **FD**
47238 **FD** 47300 **C** 47356 **FO** 47370 **FO** 47525 **IO**

LGAM Motherwell Class 56

56057 **FA** 56058 **FA** 56072 **F** 56079 **FC** 56096 **FC** 56101 **FC** 56103 **FA**
56104 **FC** 56121 **FC** 56123 **FC** 56124 **FC** 56128 **FC** 56129 **FC**

LGAY Ayr Class 08

08586 **F** 08675 **F** 08881 **D** 08906

LGBM Motherwell Class 37/0

37043 **C** 37051 **FM** 37066 **C** 37069 **C** 37071 **C** 37073 **FD** 37080 **FP**
37088 **C** 37100 **FM** 37111 **FM** 37113 **FD** 37116 **BR** 37133 **C** 37153 **C**
37154 **FD** 37165 **C** 37175 **C** 37184 **C** 37188 **C** 37196 **C** 37201 **C**
37211 **C** 37212 **FC** 37214 **FA** 37232 **C** 37240 **C** 37261 **FD** 37262 **D**
37275 37294 **C**

LGHM Motherwell Class 37/4 (West Highland)

37401 **FD** 37403 **G** 37404 **F** 37406 **FD** 37409 **F** 37410 **M** 37423 **FD**
37424 **M** 37430 **M**

LGML Motherwell Class 08

08411 08506 08622 08630 08693 08718 08720 **D**
08731 08735 08738 **D** 08882 08883 **0** 08922 **D** 08952
09103 **D** 09202 **D** 09205 **D**

LGMX Motherwell Class 08 (Stored)

08561 08938 **0**

LGPM Motherwell Class 37 HGR

37351 **C** 37692 **FC** 37693 **FC** 37696 **FC** 37712 **FP** 37714 **FM** 37801 **FC**
37893 **FP**

LGPV Motherwell Class 37/4 (ScotRail)

37427 **RR** 37428 **FP** 37431 **M**

LGSV Motherwell Class 37 (North Scotland)

37025 **BR** 37087 **C** 37099 **C** 37152 **I** 37156 **C** 37170 **C** 37221 **I**
37250 **FM** 37251 **I** 37255 **C** 37505 **I** 37510 **I** 37683 **I** 37685 **I**

LNAK Cardiff Canton Class 60 (South Wales)

60029 **FM** 60033 **FP** 60034 **FM** 60035 **FM** 60036 **FM** 60037 **FM** 60062 **FP**
60063 **FP** 60065 **FP** 60081 **FA** 60092 **FC** 60093 **FC** 60096 **FA**

LNBK Cardiff Canton Class 56 (South Wales)

56032 **FM** 56038 **FM** 56040 **FM** 56044 **FM** 56052 **F** 56053 **FM** 56060 **FM**
56064 **FM** 56073 **FM** 56076 **FM** 56113 **FC** 56114 **FC** 56115 **FC** 56119 **FC**

LNBZ St Blazey Class 08

08792 08819 **D** 08954 **F** 08955

LNCF Cardiff Canton Class 08/97

08481 08493 08576 08756 **D** 08770 **D** 08786 **D** 08798
08801 08932 08941 08942 08953 **D** 08993 08995 **FC**